PE Pipe—Design and Installation

AWWA MANUAL M55

First Edition

Science and Technology

AWWA unites the drinking water community by developing and distributing authoritative scientific and technological knowledge. Through its members, AWWA develops industry standards for products and processes that advance public health and safety. AWWA also provides quality improvement programs for water and wastewater utilities.

MANUAL OF WATER SUPPLY PRACTICES—M55, First Edition

PE Pipe—Design and Installation

Disclaimer

The authors, contributors, editors, and publisher do not assume responsibility for the validity of the content or any consequences of their use. In no event will AWWA be liable for direct, indirect, special, incidental, or consequential damages arising out of the use of information presented in this book. In particular, AWWA will not be responsible for any costs, including, but not limited to, those incurred as a result of lost revenue. In no event shall AWWA's liability exceed the amount paid for the purchase of this book.

Project Manager/Copy Editor: Melissa Christensen
Production: Glacier Publishing Services, Inc.
Manuals Coordinator: Beth Behner

Library of Congress Cataloging-in-Publication Data

PE pipe : design and installation.-- 1st ed.
p. cm.
Includes bibliographical references and index.
ISBN 1-58321-387-2
1. Pipe, Plastic Design and Construction. 2. Polyethylene.

TA448.P4 2005
628.1'5--dc22

2005055888

Printed in the United States of America.

American Water Works Association
6666 West Quincy Avenue
Denver, CO 80235-3098

ISBN 1-58321-387-2

Printed on recycled paper

Contents

This page intentionally blank.

Figures

This page intentionally blank.

Tables

Terms and Equation Symbols

Term or Symbol	Meaning
a	wave velocity (celerity), ft/sec
A	wheel contact area, in.^{2}
ATL	allowable tensile load, lb
B	float buoyancy, lb/ft
B'	soil elastic support factor
B_d	trench width at the pipe springline, in.
B_N	negative buoyancy, lb/ft
B_P	buoyancy of pipe, lb/ft
C	Hazen-Williams flow coefficient, dimensionless
D	average inside pipe diameter, ft
d	float outside diameter, in.
DF	design factor, dimensionless — the factor that is used to reduce the hydrostatic design basis to arrive at the hydrostatic design stress from which the pressure class is determined. Unless otherwise noted, the design factor for water applications is 0.5
D_i	average inside pipe diameter, in.
DIPS	ductile iron pipe size — the nominal outside diameter is the same as ductile iron pipe
D_M	mean diameter, in. ($D_o - t$)
D_o	average outside diameter of the pipe, in.
DR	dimension ratio (dimensionless) — the ratio of the average specified outside diameter to the specified minimum wall thickness (D_o/t) for outside diameter controlled polyethylene pipe
E	apparent modulus of elasticity for pipe material, psi
e	natural log base number, 2.71828
E′	design modulus of soil reaction, psi
E_d	dynamic instantaneous effective modulus of elasticity of the pipe material, psi (150,000 psi for polyethylene)
E'_E	modulus of soil reaction of embedment soil, psi
E'_N	modulus of soil reaction of native soil, psi
f	Darcy-Weisbach fraction factor, dimensionless
F	pullout force, lb
f_O	ovality compensation factor
f_{SA}	actual float submergence factor
F_T	temperature compensation multiplier, dimensionless
f_T	tensile yield design (safety) factor
f_Y	time under tension design (safety) factor
g	acceleration due to gravity, 32.2 ft/sec^2
H	soil height above pipe crown, ft
h	float submergence below water level, in.

Term or Symbol	Meaning
HDB	hydrostatic design basis, psi — the categorized long-term strength in the circumferential or hoop direction for the polyethylene material as established from long-term pressure tests in accordance with PPI TR-3 and the methodology contained in ASTM D2837
HDS	hydrostatic design stress, psi — the hydrostatic design basis multiplied by the design factor (HDB × DF)
h_f	frictional head loss, ft of liquid
H_{OT}	depth of open trench, ft
H_W	groundwater height above pipe, ft
I	moment of inertia, $in.^4$
I_C	influence coefficient, dimensionless
IDR	inside dimension ratio, dimensionless — the ratio of the average specified inside diameter to the specified minimum wall thickness (D/t) for inside diameter controlled polyethylene pipe
I_f	impact factor, dimensionless
IPS	iron pipe size — the nominal outside diameter is the same as iron (steel) pipe
K	bulk modulus of liquid at working temperature (300,000 psi for water at 73°F [23°C])
K_e	underwater environment factor
L	length of pipe, ft
L_{BS}	ballast weight spacing, ft
L_{eq}	equivalent length of straight pipe, ft — for fittings, the equivalent length of straight pipe that has the same frictional head loss as the fitting
L_F	length of float, ft
L_{OT}	length of open trench, ft
L_S	distance between supports, ft
L_{SP}	length of supported pipeline, ft
L_t	time-lag factor, dimensionless
M_M	density of foam fill, lb/ft^3
N	safety factor
P	pipe internal pressure, psi
$P_{(MAX)(OS)}$	maximum allowable system pressure during occasional surge, psi
$P_{(MAX)(RS)}$	maximum allowable system pressure during recurrent surge, psi
PC	pressure class, psi — the pressure class is the design capacity to resist working pressure up to 80°F (27°C) with specified maximum allowances for recurring positive pressure surges above working pressure. Pressure class also denotes the pipe's maximum working pressure rating for water at 80°F (27°C)
P_{CA}	allowable external pressure for constrained pipe, psi
PE	polyethylene
P_E	earth pressure on pipe, psi
PE 2406	a standard code designation for polyethylene pipe and fittings materials that has a minimum cell classification of 213333C, D, or E per ASTM D3350 and a hydrostatic design basis at 73.4°F (23°C) of 1250 psi
PE 3408	a standard code designation for polyethylene pipe and fittings materials that has a minimum cell classification of 334434C, D, or E per ASTM D3350 and a hydrostatic design basis at 73.4°F (23°C) of 1600 psi
P_{ES}	surcharge earth load pressure at point on pipe crown, psf

Term or Symbol	Meaning
P_L	vertical stress acting on pipe crown, psi
P_{OS}	pressure allowance for occasional surge pressure, psi — occasional surge pressures are caused by emergency operations that are usually the result of a malfunction such as power failure or system component failure, which includes pump seize-up, valve stem failure, and pressure-relief-valve failure
P_{RS}	pressure allowance for recurring surge pressure, psi — recurring surge pressures occur frequently and inherent in the design and operation of the system (such as normal pump startup and shutdown and normal valve opening or closure)
P_S	transient surge pressure, psi — the maximum hydraulic transient pressure increase (water hammer) in excess of the operating pressure that is anticipated in the system as a result of sudden changes in the velocity of the water column
P_{UA}	allowable external pressure for unconstrained pipe, psi
P_V	negative internal pressure (vacuum) in pipe, psi
Q	volumetric liquid flow rate, U.S. gal/min
R	equivalent radius, ft
R_b	buoyancy reduction factor
Re	Reynolds Number, dimensionless
s	hydraulic slope, ft/ft — frictional head loss per foot of pipe (h_f/L)
S	hoop compressive wall stress, psi
S_C	soil support factor
SDR	standard dimension ratio (dimensionless) — the ratio of the average specified outside diameter to the specified minimum wall thickness for outside diameter controlled polyethylene pipe, the value of which is derived by adding one to the pertinent number selected from the ANSI Preferred Number Series R10. Some of the values are as follows: R10 — SDR 5 — 6 6.3 — 7.3 8 — 9 10 — 11 12.5 — 13.5 16 — 17 20 — 21 25 — 26 31.5 — 32.5 40 — 41
S_L	specific gravity of liquid
S_P	internal pressure hoop stress, psi
t	minimum specified wall thickness, in.
t_a	average wall thickness, in. — 106% of minimum wall thickness (t*1.06)
T_Y	pipe tensile yield strength, psi
v	average velocity of flowing fluid, ft/sec
V_B	pipe bore volume, ft^3/ft
V_F	float internal volume, ft^3/ft
V_P	displaced volume of pipe, ft^3/ft
w	unit weight of soil, lb/ft^3
W	supported load, lb
W_{BD}	weight of dry ballast, lb/ft
W_{BS}	weight of submerged ballast, lb/ft

Term or Symbol	Meaning
w_F	float weight, lb/ft
WF	float load supporting capacity, lb
W_L	vehicular wheel load, lb
w_{LI}	weight of liquid inside pipe, lb/ft
w_M	weight of floam fill, lb/ft
w_P	pipe weight, lb/ft
WP	working pressure, psi — the maximum anticipated sustained operating pressure applied to the pipe exclusive of surge pressures
WPR	working pressure rating, psi — the working pressure rating is the pipe's design capacity to resist working pressure at the anticipated operating temperature with sufficient capacity against the actual anticipated positive pressure surges above working pressure. A pipe's WPR may be equal to or less than its nominal pressure class depending on the positive transient pressure characteristics of the system and pipe operating temperature if above 80°F (27°C)
W_s	distributed surcharge pressure acting over ground surface, psf
w_S	weight of float attachment structure, lb
y_S	deflection between supports, in.
γ	kinematic viscosity of the flowing fluid, ft^2/sec
Δv	velocity change occurring within the critical time 2L/a, sec
ΔY	change in diameter due to deflection, in.
ε	absolute roughness of the pipe, ft
μ	Poisson's ratio
ω_B	specific weight of ballast material, lb/ft^3
ω_L	specific weight of liquid, lb/ft^3
ω_{LI}	specific weight of the liquid inside the pipe, lb/ft^3
ω_{LO}	specific weight of the liquid outside the pipe, lb/ft^3

Conversions

METRIC CONVERSIONS

Linear Measurement

inch (in.)	× 25.4	= millimeters (mm)
inch (in.)	× 2.54	= centimeters (cm)
foot (ft)	× 304.8	= millimeters (mm)
foot (ft)	× 30.48	= centimeters (cm)
foot (ft)	× 0.3048	= meters (m)
yard (yd)	× 0.9144	= meters (m)
mile (mi)	× 1,609.3	= meters (m)
mile (mi)	× 1.6093	= kilometers (km)
millimeter (mm)	× 0.03937	= inches (in.)
centimeter (cm)	× 0.3937	= inches (in.)
meter (m)	× 39.3701	= inches (in.)
meter (m)	× 3.2808	= feet (ft)
meter (m)	× 1.0936	= yards (yd)
kilometer (km)	× 0.6214	= miles (mi)

Area Measurement

square meter (m^2)	× 10,000	= square centimeters (cm^2)
hectare (ha)	× 10,000	= square meters (m^2)
square inch ($in.^2$)	× 6.4516	= square centimeters (cm^2)
square foot (ft^2)	× 0.092903	= square meters (m^2)
square yard (yd^2)	× 0.8361	= square meters (m^2)
acre	× 0.004047	= square kilometers (km^2)
acre	× 0.4047	= hectares (ha)
square mile (mi^2)	× 2.59	= square kilometers (km^2)
square centimeter (cm^2)	× 0.16	= square inches ($in.^2$)
square meters (m^2)	× 10.7639	= square feet (ft^2)
square meters (m^2)	× 1.1960	= square yards (yd^2)
hectare (ha)	× 2.471	= acres
square kilometer (km^2)	× 247.1054	= acres
square kilometer (km^2)	× 0.3861	= square miles (mi^2)

Volume Measurement

cubic inch ($in.^3$)	× 16.3871	= cubic centimeters (cm^3)
cubic foot (ft^3)	× 28,317	= cubic centimeters (cm^3)
cubic foot (ft^3)	× 0.028317	= cubic meters (m^3)
cubic foot (ft^3)	× 28.317	= liters (L)
cubic yard (yd^3)	× 0.7646	= cubic meters (m^3)
acre foot (acre-ft)	× 123.34	= cubic meters (m^3)
ounce (US fluid) (oz)	× 0.029573	= liters (L)
quart (liquid) (qt)	× 946.9	= milliliters (mL)
quart (liquid) (qt)	× 0.9463	= liters (L)
gallon (gal)	× 3.7854	= liters (L)

gallon (gal)	× 0.0037854	= cubic meters (m^3)
peck (pk)	× 0.881	= decaliters (dL)
bushel (bu)	× 0.3524	= hectoliters (hL)
cubic centimeters (cm^3)	× 0.061	= cubic inches ($in.^3$)
cubic meter (m^3)	× 35.3183	= cubic feet (ft^3)
cubic meter (m^3)	× 1.3079	= cubic yards (yd^3)
cubic meter (m^3)	× 264.2	= gallons (gal)
cubic meter (m^3)	× 0.000811	= acre-feet (acre-ft)
liter (L)	× 1.0567	= quart (liquid) (qt)
liter (L)	× 0.264	= gallons (gal)
liter (L)	× 0.0353	= cubic feet (ft^3)
decaliter (dL)	× 2.6417	= gallons (gal)
decaliter (dL)	× 1.135	= pecks (pk)
hectoliter (hL)	× 3.531	= cubic feet (ft^3)
hectoliter (hL)	× 2.84	= bushels (bu)
hectoliter (hL)	× 0.131	= cubic yards (yd^3)
hectoliter (hL)	× 26.42	= gallons (gal)

Pressure Measurement

pound/square inch (psi)	× 6.8948	= kilopascals (kPa)
pound/square inch (psi)	× 0.00689	= pascals (Pa)
pound/square inch (psi)	× 0.070307	= kilograms/square centimeter (kg/cm^2)
pound/square foot (lb/ft^2)	× 47.8803	= pascals (Pa)
pound/square foot (lb/ft^2)	× 0.000488	= kilograms/square centimeter (kg/cm^2)
pound/square foot (lb/ft^2)	× 4.8824	= kilograms/square meter (kg/m^2)
inches of mercury	× 3,376.8	= pascals (Pa)
inches of water	× 248.84	= pascals (Pa)
bar	× 100,000	= newtons per square meter
pascals (Pa)	× 1	= newtons per square meter
pascals (Pa)	× 0.000145	= pounds/square inch (psi)
kilopascals (kPa)	× 0.145	= pounds/square inch (psi)
pascals (Pa)	× 0.000296	= inches of mercury (at 60°F)
kilogram/square centimeter (kg/cm^2)	× 14.22	= pounds/square inch (psi)
kilogram/square centimeter (kg/cm^2)	× 28.959	= inches of mercury (at 60°F)
kilogram/square meter (kg/m^2)	× 0.2048	= pounds per square foot (lb/ft^2)
centimeters of mercury	× 0.4461	= feet of water

Weight Measurement

ounce (oz)	× 28.3495	= grams (g)
pound (lb)	× 0.045359	= grams (g)
pound (lb)	× 0.4536	= kilograms (kg)
ton (short)	× 0.9072	= megagrams (metric ton)
pounds/cubic foot (lb/ft^3)	× 16.02	= grams per liter (g/L)
pounds/million gallons (lb/mil gal)	× 0.1198	= grams per cubic meter (g/m^3)
gram (g)	× 15.4324	= grains (gr)
gram (g)	× 0.0353	= ounces (oz)
gram (g)	× 0.0022	= pounds (lb)
kilograms (kg)	× 2.2046	= pounds (lb)
kilograms (kg)	× 0.0011	= tons (short)
megagram (metric ton)	× 1.1023	= tons (short)

grams/liter (g/L)	× 0.0624	= pounds per cubic foot (lb/ft^3)
grams/cubic meter (g/m^3)	× 8.3454	= pounds/million gallons (lb/mil gal)

Flow Rates

gallons/second (gps)	× 3.785	= liters per second (L/sec)
gallons/minute (gpm)	× 0.00006308	= cubic meters per second (m^3/sec)
gallons/minute (gpm)	× 0.06308	= liters per second (L/sec)
gallons/hour (gph)	× 0.003785	= cubic meters per hour (m^3/hr)
gallons/day (gpd)	× 0.000003785	= million liters per day (ML/day)
gallons/day (gpd)	× 0.003785	= cubic meters per day (m^3/day)
cubic feet/second (ft^3/sec)	× 0.028317	= cubic meters per second (m^3/sec)
cubic feet/second (ft^3/sec)	× 1,699	= liters per minute (L/min)
cubic feet/minute (ft^3/min)	× 472	= cubic centimeters/second (cm^3/sec)
cubic feet/minute (ft^3/min)	× 0.472	= liters per second (L/sec)
cubic feet/minute (ft^3/min)	× 1.6990	= cubic meters per hour (m^3/hr)
million gallons/day (mgd)	× 43.8126	= liters per second (L/sec)
million gallons/day (mgd)	× 0.003785	= cubic meters per day (m^3/day)
million gallons/day (mgd)	× 0.043813	= cubic meters per second (m^3/sec)
gallons/square foot (gal/ft^2)	× 40.74	= liters per square meter (L/m^2)
gallons/acre/day (gal/acre/day)	× 0.0094	= cubic meters/hectare/day (m^3/ha/day)
gallons/square foot/day (gal/ft^2/day)	× 0.0407	= cubic meters/square meter/day (m^3/m^2/day)
gallons/square foot/day (gal/ft^2/day)	× 0.0283	= liters/square meter/day (L/m^2/day)
gallons/square foot/minute (gal/ft^2/min)	× 2.444	= cubic meters/square meter/hour (m^3/m^2/hr) = m/hr
gallons/square foot/minute (gal/ft^2/min)	× 0.679	= liters/square meter/second (L/m^2/sec)
gallons/square foot/minute (gal/ft^2/min)	× 40.7458	= liters/square meter/minute (L/m^2/min)
gallons/capita/day (gpcd)	× 3.785	= liters/day/capita (L/d per capita)
liters/second (L/sec)	× 22,824.5	= gallons per day (gpd)
liters/second (L/sec)	× 0.0228	= million gallons per day (mgd)
liters/second (L/sec)	× 15.8508	= gallons per minute (gpm)
liters/second (L/sec)	× 2.119	= cubic feet per minute (ft^3/min)
liters/minute (L/min)	× 0.0005886	= cubic feet per second (ft^3/sec)
cubic centimeters/second (cm^3/sec)	× 0.0021	= cubic feet per minute (ft^3/min)
cubic meters/second (m^3/sec)	× 35.3147	= cubic feet per second (ft^3/sec)
cubic meters/second (m^3/sec)	× 22.8245	= million gallons per day (mgd)
cubic meters/second (m^3/sec)	× 15,850.3	= gallons per minute (gpm)
cubic meters/hour (m^3/hr)	× 0.5886	= cubic feet per minute (ft^3/min)
cubic meters/hour (m^3/hr)	× 4.403	= gallons per minute (gpm)
cubic meters/day (m^3/day)	× 264.1720	= gallons per day (gpd)
cubic meters/day (m^3/day)	× 0.00026417	= million gallons per day (mgd)
cubic meters/hectare/day (m^3/ha/day)	× 106.9064	= gallons per acre per day (gal/acre/day)
cubic meters/square meter/day (m^3/m^2/day)	× 24.5424	= gallons/square foot/day (gal/ft^2/day)
liters/square meter/minute (L/m^2/min)	× 0.0245	= gallons/square foot/minute (gal/ft^2/min)
liters/square meter/minute (L/m^2/min)	× 35.3420	= gallons/square foot/day (gal/ft^2/day)

Work, Heat, and Energy

British thermal units (Btu)	× 1.0551	= kilojoules (kJ)
British thermal units (Btu)	× 0.2520	= kilogram-calories (kg-cal)
foot-pound (force) (ft-lb)	× 1.3558	= joules (J)
horsepower-hour (hp·hr)	× 2.6845	= megajoules (MJ)
watt-second (W-sec)	× 1.000	= joules (J)
watt-hour (W·hr)	× 3.600	= kilojoules (kJ)
kilowatt-hour (kW·hr)	× 3,600	= kilojoules (kJ)
kilowatt-hour (kW·hr)	× 3,600,000	= joules (J)
British thermal units per pound (Btu/lb)	× 0.5555	= kilogram-calories per kilogram (kg-cal/kg)
British thermal units per cubic foot (Btu/ft^3)	× 8.8987	= kilogram-calories/cubic meter (kg-cal/m^3)
kilojoule (kJ)	× 0.9478	= British thermal units (Btu)
kilojoule (kJ)	× 0.00027778	= kilowatt-hours (kW·hr)
kilojoule (kJ)	× 0.2778	= watt-hours (W·hr)
joule (J)	× 0.7376	= foot-pounds (ft-lb)
joule (J)	× 1.0000	= watt-seconds (W-sec)
joule (J)	× 0.2399	= calories (cal)
megajoule (MJ)	× 0.3725	= horsepower-hour (hp·hr)
kilogram-calories (kg-cal)	× 3.9685	= British thermal units (Btu)
kilogram-calories per kilogram (kg-cal/kg)	× 1.8000	= British thermal units per pound (Btu/lb)
kilogram-calories per liter (kg-cal/L)	× 112.37	= British thermal units per cubic foot (Btu/ft^3)
kilogram-calories/cubic meter (kg-cal/m^3)	× 0.1124	= British thermal units per cubic foot (Btu/ft^3)

Velocity, Acceleration, and Force

feet per minute (ft/min)	× 18.2880	= meters per hour (m/hr)
feet per hour (ft/hr)	× 0.3048	= meters per hour (m/hr)
miles per hour (mph)	× 44.7	= centimeters per second (cm/sec)
miles per hour (mph)	× 26.82	= meters per minute (m/min)
miles per hour (mph)	× 1.609	= kilometers per hour (km/hr)
feet/second/second (ft/sec^2)	× 0.3048	= meters/second/second (m/sec^2)
inches/second/second (in./sec^2)	× 0.0254	= meters/second/second (m/sec^2)
pounds force (lbf)	× 4.44482	= newtons (N)
centimeters/second (cm/sec)	× 0.0224	= miles per hour (mph)
meters/second (m/sec)	× 3.2808	= feet per second (ft/sec)
meters/minute (m/min)	× 0.0373	= miles per hour (mph)
meters per hour (m/hr)	× 0.0547	= feet per minute (ft/min)
meters per hour (m/hr)	× 3.2808	= feet per hour (ft/hr)
kilometers/second (km/sec)	× 2.2369	= miles per hour (mph)
kilometers/hour (km/hr)	× 0.0103	= miles per hour (mph)
meters/second/second (m/sec^2)	× 3.2808	= feet/second/second (ft/sec^2)
meters/second/second (m/sec^2)	× 39.3701	= inches/second/second (in./sec^2)
newtons (N)	× 0.2248	= pounds force (lbf)

CELSIUS/FAHRENHEIT COMPARISON GRAPH

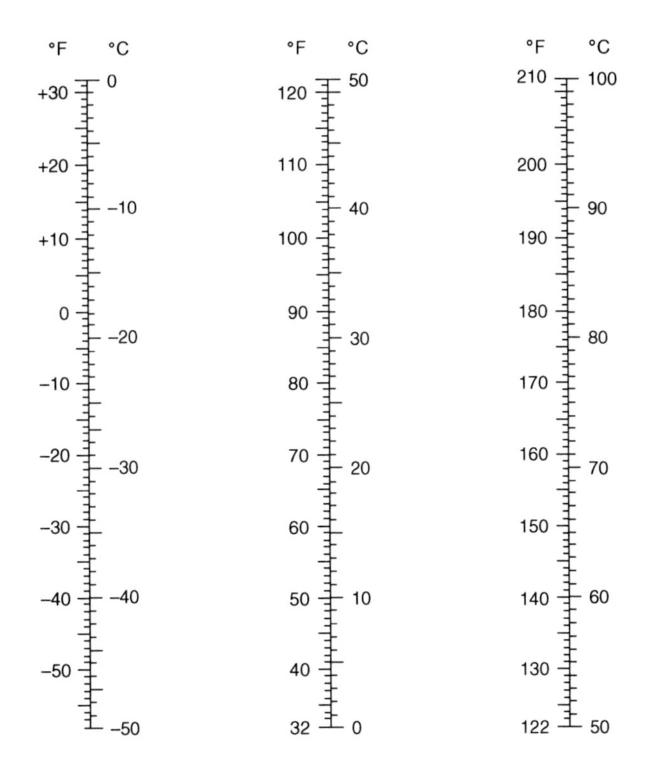

0.555 (°F – 32)	=	degrees Celsius (°C)
(1.8 × °C) + 32	=	degrees Fahrenheit (°F)
°C + 273.15	=	kelvin (K)
boiling point*	=	212 °F
	=	100 °C
	=	373 K
freezing point*	=	32 °F
	=	0 °C
	=	273 K

*At 14.696 psia, 101.325 kPa.

DECIMAL EQUIVALENTS OF FRACTIONS

Fraction	Decimal	Fraction	Decimal
1/64	0.01563	33/64	0.51563
1/32	0.03125	17/32	0.53125
3/64	0.04688	35/64	0.54688
1/16	0.06250	9/16	0.56250
5/64	0.07813	37/64	0.57813
3/32	0.09375	19/32	0.59375
7/64	0.10938	39/64	0.60938
1/8	0.12500	5/8	0.62500
9/64	0.14063	41/64	0.64063
5/32	0.15625	21/32	0.65625
11/64	0.17188	43/64	0.67188
3/16	0.18750	11/16	0.68750
13/64	0.20313	45/64	0.70313
7/32	0.21875	23/32	0.71875
15/64	0.23438	47/64	0.73438
1/4	0.25000	3/4	0.75000
17/64	0.26563	49/64	0.76563
9/32	0.28125	25/32	0.78125
19/64	0.29688	51/64	0.79688
10/32	0.31250	13/16	0.81250
21/64	0.32813	53/64	0.82813
11/32	0.34375	27/32	0.84375
23/64	0.35938	55/64	0.85938
3/8	0.37500	7/8	0.87500
25/64	0.39063	57/64	0.89063
13/32	0.40625	29/32	0.90625
27/64	0.42188	59/64	0.92188
7/16	0.43750	15/16	0.93750
29/64	0.45313	61/64	0.95313
15/32	0.46875	31/32	0.96875
31/64	0.48438	63/64	0.98438
1/2	0.50000		

Preface

This is the first edition of AWWA M55 *PE Pipe—Design and Installation*. The manual provides the user with both technical and general information to aid in the design, specification, procurement, installation, and understanding of the high-density polyethylene (HDPE) pipe and fittings. It is a discussion of recommended practice, not an American Water Works Association (AWWA) standard calling for compliance with certain specifications. It is intended for use by utilities and municipalities of all sizes, whether as a reference book or textbook for those not fully familiar with HDPE pipe and fittings products. Municipal and consulting engineers may use this manual in preparing plans and specifications for new HDPE pipe projects.

The manual describes HDPE pipe and fittings products and certain appurtenances, and their applications to practical installations, whether of a standard or special nature. For adequate knowledge of these products, the entire manual needs to be studied. Readers will also find the manual a useful source of information for assistance with specific or unusual conditions. The manual contains a list of applicable national standards, which may be purchased from the respective standards organizations (e.g., AWWA, ASTM, etc.). Readers should use the latest editions of the Standards that are referenced.

Credit is extended to The Plastics Pipe Institute, Inc. (www.plasticpipe.org) for its contribution to the manual.

This page intentionally blank.

Acknowledgments

The following members of the PE Manual Subcommittee and the Polyolefin Pressure Pipe and Fittings Committee helped author this new manual.

William I. Adams, W.L. Plastics, Cedar City, Utah
Will Bezner, CSR Poly Pipe Inc., Gainesville, Texas
Dudley Burwell, ISCO Industries, Huntsville, Ala.
Nancy Conley, NOVA Chemicals, Kirkland, Que.
Jim M. Craig, McElroy Manufacturing Inc., Tulsa, Okla.
Richard P. Fuerst, Bureau of Reclamation, Denver, Colo.
Larry J. Petroff, Performance Pipe, Plano, Texas
Steve D. Sandstrum, BP Solvay Polyethylene North America, Deer Park, Texas
Terry Stiles, Central Plastics Company, Shawnee, Okla.

This new manual was reviewed and approved by the PE Manual Subcommittee and the Polyolefin Pressure Pipe and Fittings Committee and included the following personnel through the time of development and approval:

Camille G. Rubeiz, *Chair*

W.I. Adams, W.L. Plastics, Cedar City, Utah
Will Bezner, CSR Poly Pipe Inc., Gainesville, Texas
M.G. Boyle, Pflugerville, Texas
Dudley Burwell, ISCO Industries, Huntsville, Ala.
Nancy Conley, NOVA Chemicals, Kirkland, Que.
J.D. Cox, Stockton, Calif.
J.M. Craig, McElroy Manufacturing Inc., Tulsa, Okla.
R.P. Fuerst, Bureau of Reclamation, Denver, Colo.
S.W. King, North American Pipe Corporation, Houston, Texas
L.J. Petroff, Performance Pipe, Plano, Texas
C.G. Rubeiz, Plastics Pipe Institute, Washington, D.C.
S.D. Sandstrum, BP Solvay Polyethylene North America, Deer Park, Texas
Terry Stiles, Central Plastics Company, Shawnee, Okla.
Donna Stoughton, Charter Plastics Inc., Titusville, Pa.
Harvey Svetlik, Independent Pipe Products Inc., Dallas, Texas

Michael G. Boyle, *Chair*

General Interest Members

J.P. Castronovo, CH2M Hill, Gainesville, Fla.
K.C. Choquette, Des Moines, Iowa
*W.J. Dixon**, Dixon Engineering Inc., Lake Odessa, Mich.
D.E. Duvall, Engineering Systems Inc., Aurora, Ill.

---

* Liaison

---

* Liaison

Chapter 1

Engineering Properties of Polyethylene

INTRODUCTION

A fundamental understanding of material characteristics is an inherent part of the design process for any piping system. With such an understanding, the piping designer can use the properties of the material to design for optimum performance. This chapter provides basic information that should assist the reader in understanding how polyethylene's (PE's) material characteristics influence its engineering behavior.

PE is a thermoplastic, which means that it is a polymeric material that can be softened and formed into useful shapes by the application of heat and pressure and which hardens when cooled. PE is a member of the polyolefins family, which also includes polypropylene. As a group of materials, the polyolefins generally possess low water absorption, moderate to low gas permeability, good toughness and flexibility at low temperatures, and a relatively low heat resistance. PE plastics form flexible but tough products and possess excellent resistance to many chemicals.

POLYMER CHARACTERISTICS

In general terms, the performance capability of PE in piping applications is determined by three main parameters: density, molecular weight, and molecular weight distribution. Each of these polymer properties has an effect on the physical performance associated with a specific PE resin. The general effect of variation in these three physical properties as related to polymer performance is shown in Table 1-1.

Density

PE is a semicrystalline polymer composed of long, chain-like molecules of varying lengths and numbers of side branches. As the number of side branches increases, polymer crystallinity and hence, density decreases because the molecules cannot pack as

Table 1-1 Effects of density, molecular weight, and molecular weight distribution

Property	As Density Increases	As Molecular Weight Increases	As Molecular Weight Distribution Broadens
Tensile	Increases	Increases	—
Stiffness	Increases	Increases slightly	Decreases
Impact strength	Decreases	Increases	Decreases
Low temperature brittleness	Increases	Decreases	Decreases
Abrasion resistance	Increases	Increases	—
Hardness	Increases	Increases slightly	—
Softening point	Increases	—	Increases
Stress crack resistance	Decreases	Increases	Increases
Permeability	Decreases	Increases slightly	—
Chemical resistance	Increases	Increases	—
Melt strength	—	Increases	Increases

closely together. Density affects many of the physical properties associated with the performance of the finished pipe. Properties such as stress crack resistance, tensile strength, and stiffness are all affected by the base resin density of the polymer as shown in Table 1-1.

Base resin density refers to the density of the natural PE that has not been compounded with additives and/or colorants. Within this range, the materials are generically referred to as either medium or high density in nature. PE pipe resins with a base resin density in the range of 0.935 to 0.941 grams per cubic centimeter (g/cc) are referred to as medium density PE. PE pipe base resins in the range of 0.941 to 0.945 g/cc are commonly referred to as high-density polyethylenes (HDPEs). Industry practice has shown that base resin (unpigmented) densities in the range of 0.936 to 0.945 g/cc offer a highly beneficial combination of performance properties for the majority of piping applications.

The addition of carbon black to the base PE resin does have an impact on the compounded density of the material. The addition of 2 to 2.5 percent carbon black raises the compounded material density on the order of 0.009–0.011 g/cc. The variability in the actual percentage of carbon black incorporated can have a moderate affect on comparative density values. As a result, industry practice as established by ASTM standard is to provide comparative values on the base resin density as this is a better indicator of the polymer crystallinity.

Molecular Weight

PE resins are composed of a number of molecular chains of varying lengths. As a result, the molecular weight of the resin is the average of the weight of each of these chains. The average weight may be determined using sophisticated scientific techniques, such as gel permeation chromatography or size-exclusion chromatography. For PE of a given density, the effect of increasing molecular weight on physical properties is shown in Table 1-1.

A very rough indicator of the molecular weight of a polymer may be obtained using the melt index technique of analysis as described in ASTM D1238[1]. The melt index technique is an inexpensive means of comparing, in a relative manner, the molecular weight of PEs having similar structure. Resins with a relatively low average molecular weight will have a comparatively high melt index. Conversely, resins with a relatively high molecular weight will yield a lower melt index. From this relationship, we can associate changes in physical properties (as shown in Table 1-1) with changes in melt index of the material. It is important not to use melt index alone as a definitive indicator of molecular weight because variations in polymer structure can affect both molecular weight and melt index.

Molecular Weight Distribution

Molecular weight distribution (MWD) refers to the statistical grouping of the individual molecular chains within a PE resin. Resins made up of molecules that vary considerably in molecular weight are considered to have a broad MWD. When most of the molecules are nearly the same length, the MWD is considered narrow. The effect of broadening the MWD of a PE resin having a given density and molecular weight is shown in Table 1-1.

Recent Advances

It should be noted that recent advances in polymer technology have led to the development and introduction of even higher density resins for use in piping applications. These new materials that have base resin densities as high as 0.952 g/cc in combination with higher molecular weight and bimodal molecular weight distribution are generally recognized as offering higher levels of technical performance under ISO standards for PE piping that are common outside of North America. These higher levels of technical performance are not yet recognized within the North American standards system.

MECHANICAL PROPERTIES

Viscoelasticity

PE is characterized as a viscoelastic construction material. Because of its molecular nature, PE is a complex combination of elastic-like and fluid-like elements. As a result, this material displays properties that are intermediate to crystalline metals and very high viscosity fluids. Figure 1-1 is the traditional diagrammatic representation of PE in which the springs represent those components of the PE matrix that respond to loading in a traditional elastic manner in accordance with Hooke's law. The dashpots represent fluid elements of the polymer that respond to load much as a Newtonian fluid.

As a result of the viscoelastic character of the polymer, the tensile stress–strain curve for PE is divided into three distinct regions. The first of these is an initial linear deformation in response to the load imposed that is generally recoverable when the load is removed. In the second stage of loading, deformation continues but at an ever decreasing rate. Thus, the slope of the stress–strain curve is constantly changing, attesting to its curvilinear nature. Deformation in the second stage may not be fully recoverable. The final stage of the stress–strain curve for PE is characterized by necking down followed by distinct elongation or extension ultimately ending in ductile rupture of the material.

The viscoelastic nature of PE provides for two unique engineering characteristics that are employed in the design of HDPE water piping systems. These are *creep* and *stress relaxation*.

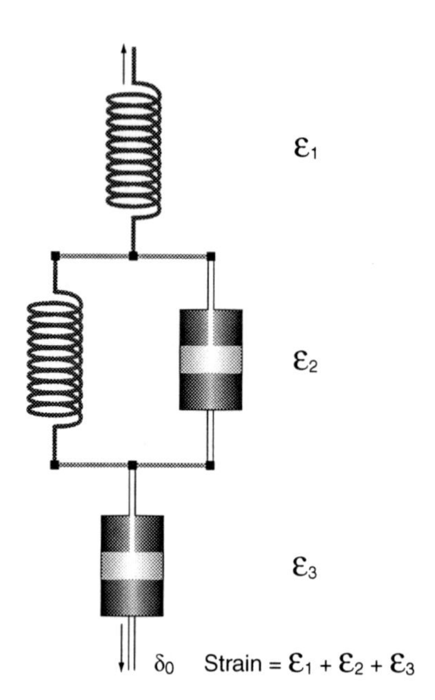

Figure 1-1 Traditional model of HDPE

Creep is not an engineering concern as it relates to PE piping materials. Creep refers to the response of PE, over time, to a constant static load. When HDPE is subjected to a constant static load, it deforms immediately to a strain predicted by the stress–strain modulus determined from the tensile stress–strain curve. The material continues to deform indefinitely at an ever decreasing rate. If the load is high enough, the material may yield or rupture. This time-dependent viscous flow component of deformation is called creep. This asserts that the long-term properties of PE are not adequately predicted by the results of short-term testing, such as tensile strength. As such, PE piping materials are designed in accordance with longer-term tests such as hydrostatic testing and testing for resistance to slow crack growth, which when used in accordance with industry recommended practice, the resultant deformation caused by sustained loading, or creep, is not sufficiently large to be an engineering concern.

Stress relaxation is another unique property arising from the viscoelastic nature of PE. When subjected to a constant strain (deformation of a specific degree) that is maintained over time, the load or stress generated by the deformation slowly decreases over time. This is of considerable importance to the design of PE piping systems.

Because of its viscoelastic nature, the response of PE piping systems to loading is time-dependent. The effective modulus of elasticity is significantly reduced by the duration of the loading because of the creep and stress relaxation characteristics of PE. An instantaneous modulus for sudden events such as water hammer can be as high as 150,000 psi at 73°F (23°C). For slightly longer duration, but short-term events such as soil settlement and live loadings, the effective modulus for PE is roughly 110,000 to 120,000 psi at 73°F (23°C), and as a long-term property, the effective long-term modulus calculates to be approximately 20,000 to 38,000 psi. This modulus becomes the criteria for the long-term design life of PE piping systems.

This same time-dependent response to loading is also what gives PE its unique resiliency and resistance to sudden, comparatively short-term loading phenomena.

Such is the case with PE's resistance to water hammer, which will be discussed in more detail in subsequent sections.

PE is a thermoplastic and, as such, its properties are temperature dependent as well as dependent on the duration of loading. Therefore, the absolute value of the engineering properties of PE will vary in relation to the temperature at which the specific tests are conducted. Industry convention is to design PE piping systems using engineering properties established at the standard temperature of 73°F (23°C) and then employ industry established temperature compensating multipliers to provide for the service condition temperatures.

Tensile Strength

Tensile strength is a short-term property that provides a basis for classification or comparison when established at specific conditions of temperature and rate of loading but is of limited significance from a design perspective. The tensile strength of PE is typically determined in accordance with ASTM D638[2]. In this test, PE specimens are prepared and pulled in a controlled environment at a constant rate of strain.

Any material will deform when a force is applied. The amount of deformation per unit length is termed the *strain,* and the force per cross-sectional area is termed the *stress.* As it relates to tensile testing of PE pipe grades, strain is generally approximated by assuming a straight-line relationship to stress at lower stress levels (up to 30 percent of the tensile yield point), and it is reversible. That is, the material deforms but will over time recover its original shape when the stress is removed. The strain in this region is referred to as the elastic strain because it is reversible. The Modulus of Elasticity (or Young's Modulus) is the ratio between the stress and strain in this reversible region.

At stress levels generally greater than 50 percent, strain is no longer proportional to stress and is not reversible, that is, the slope of the stress–strain curve changes at an increasing rate. At these higher stress levels, the materials begin to deform such that the original dimensions are not recoverable. In actual testing of PE pipe grade materials, this stage is characterized by initiation of a distinct "necking" of the tensile specimen. This is called the *plastic strain region.* The point at which stress causes a material to deform beyond the elastic region is termed the *tensile strength at yield.* The stress required to ultimately break the test specimen is called the *ultimate tensile strength* or the *tensile strength at break.* (See Figure 1-2.)

Of equal importance is the percent elongation obtained during tensile testing because this information can provide a relative indication of the ductility of the polymer being evaluated. Materials with relatively high levels of elongation are indicative of highly ductile performance as pipe. Modern pipe grade PEs will demonstrate elongations of 400 to 800 percent or more between yield and ultimate tensile rupture. It is also typical that tensile strength at yield and tensile strength at break are similar values; that is, once the material yields, the load required to continue specimen elongation and eventually break the specimen changes very little.

Compressive Properties

Compressive forces act in the opposite direction to tensile forces. The effect of compressive force on PE can be measured on a tensile test apparatus using the protocol described in ASTM D6953. At small strains (up to 3 percent for most PE pipe resins), the compressive modulus is about equal to the elastic modulus. However, unlike tensile loading, which can result in a failure, compression produces a slow and infinite yielding that seldom leads to a failure. For this reason, it is customary to report compressive

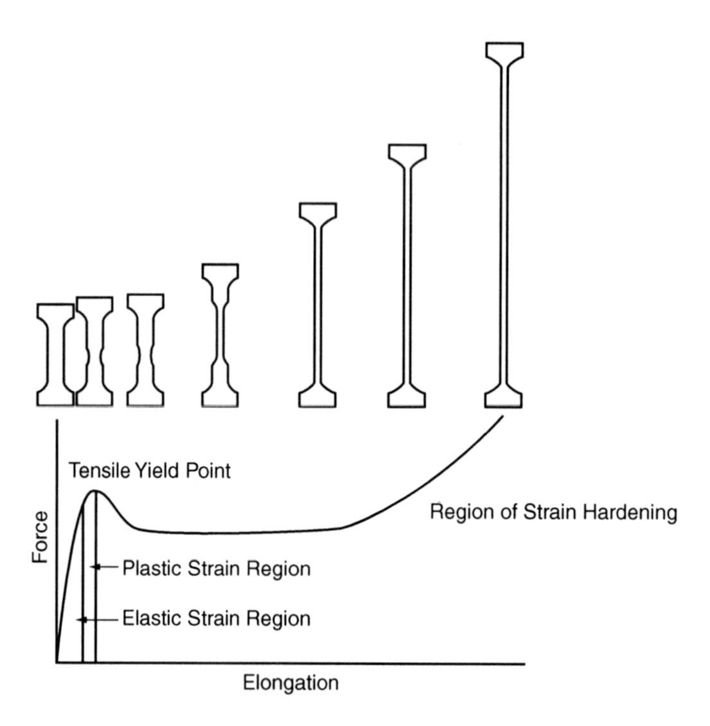

Figure 1-2 Generalized tensile stress–strain curve for PE pipe grades

strength as the stress required to deform the test specimen to a specific strain. Under conditions of mild compression, the general engineering assumption is that the effective compressive modulus is essentially equivalent to the effective tensile modulus.

Flexural Properties

The flexural strength of a material is the maximum stress in the outer fiber of a test specimen at rupture. Because most PE pipe resins do not break under this test, the true flexural strength of these materials cannot be determined. As such, the flexural modulus is typically calculated on the basis of the amount of stress required to obtain a 2 percent strain in the outer fiber. The prevailing test method is ASTM D790[4]. Depending on the density of the base resin, the effective flexural modulus of PE can range from 80,000 to 160,000 psi. The flexural modulus of PE is a short-term property that provides a basis for classification but is of limited significance from a design perspective.

Impact Properties

The amount of energy that a material can absorb without breaking or fracturing is referred to as the impact strength of that material. ASTM D256[5] describes the two most commonly used tests for PE pipe compounds, the Izod Impact Test and the Charpy Impact Test. Both test methods measure the ability of a PE specimen to absorb energy on failure. Obviously, test information such as this is used to make a relative comparison of the material's resistance to failure on impact under defined circumstances. In this regard, PE is a very tough material demonstrating Izod impact resistance values in the range of 10–12 ft-lbf/in. at standard room temperature. This is the range in which PE pipe grades will bend or deflect in response to Izod impact testing. These values will change to some degree as the temperature at which the test is conducted changes. When Izod impact testing is conducted at very low temperature (< 0°F), fracture may occur.

Abrasion Resistance

PE demonstrates outstanding abrasion resistance under potable water flow conditions. Moreover, the abrasion resistant nature of this material has resulted in the widespread use of PE pipe for liquid slurry handling applications. However, the factors that affect the wear resistance of liquid slurry pipelines are diverse. In addition to flow velocity, one must consider the type of flow regime: laminar (single phase or double phase) or turbulent flow; presence, size, angularity, and concentration of suspended solids; and angle of impingement. While these factors are germane to slurry handling applications, they will have little or no effect on the abrasion resistance of PE pipe used in the transport of clean potable water. At higher flow velocities typical of potable water distribution, there is no erosional effect on PE pipe.

OTHER PHYSICAL PROPERTIES

Permeability

The rate of transmission of gases and vapors through polymeric materials varies with the structure of both the permeating molecules and the polymer. Permeability is directly related to the crystallinity of the PE and the size and polarity of the molecule attempting to permeate through the matrix. The higher the crystallinity (the higher the density), the more resistant is the polymer to permeation. PE resins used for the manufacture of water pipe in accordance with ANSI/AWWA C906[6] possess density ranges that make them highly resistant to most types of permeation.

The designer should be aware, however, that all piping systems are susceptible to permeation of light hydrocarbon contaminants that may be present in the soil. With continued exposure over time, these contaminants can permeate from the soil into the pipe itself either through the wall of a plastic pipe or through the elastomeric gasketed joint of a mechanically joined piping system. For this reason, special care should be taken when installing potable water lines through contaminated soils regardless of the type of pipe material (concrete, plastic, ductile iron, etc.).

From ANSI/AWWA C906, Sec. 4.1:

> "The selection of materials is critical for water service and distribution piping in locations where the pipe will be exposed to significant concentrations of pollutants comprised of low molecular weight petroleum products or organic solvents or their vapors. Research has documented that pipe materials, such as PE, polybutylene, polyvinyl chloride, and asbestos cement and elastomers, such as used in jointing gaskets and packing glands, are subject to permeation by lower molecular weight organic solvents or petroleum products. If a water pipe must pass through a contaminated area or an area subject to contamination, consult with pipe manufacturers regarding permeation of pipe walls, jointing materials, etc., before selecting materials for use in that area."

Temperature Effects

PE is a thermoplastic polymer. As such, its physical properties change in response to temperature. These property changes are reversible as the temperature fluctuates. The physical properties of PE are normally determined and published at standard laboratory conditions of 73°F (23°C) with the understanding that the absolute values may change in response to temperature.

For example, the pressure rating of a PE pipe relates directly to the hydrostatic design basis (HDB) of the material from which it is produced. Traditionally, this design property is established at 73°F (23°C). However, as temperature increases, the viscoelastic nature of the polymer yields a lower modulus of elasticity, lower tensile strength, and lower stiffness. As a result, the hydrostatic strength of the material decreases, which yields a lower pressure rating for a specific pipe DR. The effect is reversible in that once the temperature decreases again to standard condition, the pressure capability of the product returns to its normal design basis. However, the elevated temperature pressure rating is always applied for elevated temperature service conditions.

Buried potable water systems typically operate in a range below 73°F (23°C). In these situations, the pressure capability of the pipe may actually exceed the design pressure class ratings listed in ANSI/AWWA C901 and C906. The current industry practice is to set the pressure rating of the pipe at 73°F (23°C) as the standard and consider any added strength at lower service temperatures as an additional factor of safety for design purposes.

The coefficient of linear expansion for unrestrained PE is generally accepted to be 1.2×10^{-4} in./in./°F. This suggests that unrestrained PE will expand or contract considerably in response to thermal fluctuation. It should be pointed out, however, that while the coefficient of expansion for PE is fairly high compared to metal piping products, the modulus of elasticity is comparatively low, approximately $^{1}/_{300}$ of steel for example. This suggests that the tensile or compressive stresses associated with a temperature change are comparatively low and can be addressed in the design and installation of the piping system. Thermal expansion and contraction effects must be taken into account for surface, above grade, and marine applications where pipe restraint may be limited. But with buried installations, soil friction frequently provides considerable restraint against thermal expansion and contraction movement. In smaller diameter installation, such as those less than 12-in. nominal outside diameter, soil friction restraint can be enhanced by snaking the pipe side to side in the trench prior to backfilling. Additional restraint against movement can be provided with in-line anchors. (See Chapter 8.)

In consideration of its thermal properties, PE pipe must be joined using methods that provide longitudinal thrust restraint such as heat fusion, electrofusion, flange connections, and restrained mechanical connections. Additionally, fittings used within the system should possess sufficient pull-out resistance in light of anticipated movement caused by thermal expansion or contraction. Finally, PE pipe should be stabilized or anchored at its termination points to other, more rigid piping or appurtenances to avoid potential stress concentration at the point of transition or to avoid excessive bending moments on system fittings. The reader is referred to Chapter 8 of this manual for more information regarding control of pull-out forces.

Electrical Properties

PE is an excellent insulator and does not conduct electricity. The typical electrical properties of PE are shown in Table 1-2.

Table 1-2 Electrical properties of PE

Electrical Property	Units	Test Method	Value
Volume Resistivity	ohms-cm	ASTM D257	$> 10^{16}$
Surface Resistivity	ohms	ASTM D257	$> 10^{13}$
Arc Resistance	seconds	ASTM D495	200 to 250
Dielectric Strength	volts/mil	ASTM D149	450 to 1,000
Dielectric Constant	—	ASTM D150	2.25 to 2.35 @ 60 Hz
Dissipation Factor	—	ASTM D150	> 0.0005 @ 60 Hz

CHEMICAL PROPERTIES

Chemical Resistance

An integral part of any piping system design is the assessment of the chemical environment to which the piping will be exposed and the impact it may have on the design life of the pipe. Generally, PE is widely recognized for its unique chemical resistance. As such, this piping material has found extensive utilization in the transport of a variety of aggressive chemicals.

To assist the designer in the selection of PE for piping applications, chemical resistance charts have been published that provide some basic guidelines regarding the suitability of PE as a piping material in the presence of various chemicals. A very comprehensive chemical resistance chart has been published by the Plastics Pipe Institute (PPI) in the *Handbook of Polyethylene Pipe*[7].

It is important to note that chemical resistance tables are only a guideline. Data such as this is generally developed on the basis of laboratory tests involving the evaluation of tensile coupons immersed in various concentrations of the reference chemicals. As such, these charts provide a relative indication of the suitability of PE when exposed. They do not assess the impact that continual exposure to these chemicals may have on various aspects of long-term performance nor do they address the effect produced by exposure to various combinations of the chemicals listed. Additionally, these chemical resistance tables do not take into consideration the affect of stress (loading), magnitude of the stress, or duration of application of such stress. In light of this, it is recommended that the designer use responsible judgment in the interpretation of this type of data and its utilization for design purposes. Additional information is available from PPI Technical Report TR-19[8]. Alternatively, the reader is referred to the pipe manufacturer who may have actual field experience under similar specific service conditions.

Corrosion

PE used in water piping applications is an electrically nonconductive polymer and not adversely affected by naturally occurring soil conditions. As such, it is not subject to galvanic action and does not rust or corrode. This aspect of PE pipe means that cathodic protection is not required to protect the long-term integrity of the pipe even in the most corrosive environments. Proper consideration should be given to any metal fittings that may be used to join the pipe or system components.

Tuberculation

The potential for tuberculation of PE pipe is minimal. Tuberculation typically occurs in response to the deposition of soluble encrustants onto the surface of the pipe and subsequent corrosive action with the base material of the pipe. Properly extruded, PE pipe has an extremely smooth surface, which provides minimal opportunity for the precipitation of minerals such as calcium carbonate and the like onto the interior surface. PE itself is inert and therefore not prone to galvanic action, which these solubles may initiate in other piping materials.

Resistance to Slow Crack Growth

PE piping manufactured in accordance with the requirements of ANSI/AWWA C901 or C906 is resistant to slow crack growth when used in typical potable water systems. Research in the area of slow crack growth combined with continual advancements in material science have resulted in HDPE piping products that when manufactured and installed in accordance with these standards are designed to provide sustained resistance to slow crack growth phenomena such as environmental stress cracking. To understand the significance of this statement, one must first understand the nature of slow crack growth and pipe failure in general.

Excluding third party damage phenomena, such as dig-ins, etc., pipe failure may occur in one of three ways. First is the sudden yielding of the pipe profile in response to a stress level beyond the design capability of the material itself. Generally, this is referred to as Stage I type failure and is typically ductile-mechanical in nature and appearance. The pressure class designations and working pressure-rating methodology presented in ANSI/AWWA C906 are developed within the constraints of these material capabilities. The material requirements stipulated in ANSI/AWWA C906 combined with additional pipe requirements, such as workmanship, dimensional specifications for each pressure class, and the five-second pressure test, provide a basis for resistance to this type of failure over the design life of the PE piping system.

The second mode of pipe failure is the result of slow crack growth. Generally, this is referred to as Stage II brittle-mechanical type failure. In this mode, pipe failure is characterized by very small slit-type failures in the pipe wall that initiate at points of mechanical stress concentration associated with inhomogeneities in the pipe wall or at imperfections on the inner pipe surface. Typically, these types of failures are slower in nature and occur as a three-stage process: crack initiation, crack propagation, and final ligament yield that results in pipe failure. This type of failure phenomena may be the result of exposure to more aggressive conditions such as elevated temperature (> 140°F [60°C]) or the oxidation reduction potential (ORP) of the water system, which is a function of chemical concentration (chlorine, chloramines, chlorine dioxide, ozone, dissolved oxygen, etc.) or other factors that are not typical of the majority of potable water applications. ANSI/AWWA C906 places specific requirements on the pipe manufactured in accordance with this standard to guard against Stage II type failures while in potable water service.

ANSI/AWWA C906 requires that all pipe must be produced from a material for which a PPI hydrostatic design basis (HDB) has been recommended. This requirement ensures that stress-rupture data for pipe specimens produced from the listed material is reviewed in accordance with the protocol in PPI's TR-3 to ensure that it meets the stress-rating requirements of ASTM D2837. The stress-rupture data is further analyzed to ensure that it "validates." That is, additional higher temperature stress-rupture tests are conducted to validate that the slope of the regression curve obtained at a specific temperature does not change until some time after the 100,000-hour requirement established within ASTM D2837. Second, ANSI/AWWA C906 also has specific performance requirements for the manufactured pipe or fittings such as thermal stability, the elevated-temperature sustained-pressure test, and the bend back test, which minimize the potential for Type II failures in typical potable water service applications.

As a safeguard against Type II failure phenomenon, piping products manufactured in accordance with ANSI/AWWA C906 are produced from PE resins that are highly resistant to environmental stress cracking as determined by the tests described below.

Laboratory tests to assess resistance to environmental stress cracking include ASTM D1693[9] and ASTM F1473[10]. These standard test methods are utilized within the plastic pipe industry to assess the piping material's resistance to cracking under accelerated conditions of concentrated stress, aggressive chemical attack, and elevated temperature. According to ASTM D1693, 10 compression molded specimens of the PE material are prepared, deformed into a 180° U-bend, and submerged in an aggressive stress-cracking chemical such as Igepal CO630 (a strong detergent) at 100°C. The specimens are maintained at elevated temperature and the time to failure is recorded. Failure is defined as cracks that are visible on the surface of the specimens.

Because ASTM D1693 defines the time to the appearance of cracks on the surface of the material, it provides information about the material's resistance to the initiation of stress cracks. Modern PE pressure piping materials have been formulated and engineered to provide excellent resistance to the initiation of stress cracks. When tested in accordance with ASTM D1693, specimens commonly do not fail in thousands of hours. More recently, PE pressure piping materials have been developed to resist stress crack initiation to such an extent that they now cannot be adequately characterized by ASTM D1693. As a result, new tests have been developed that assess the slow crack growth resistance of the materials. Predominant among these tests is ASTM F1473, which like ASTM D1693 has been incorporated into ASTM D3350 as a classified slow crack growth resistance property.

ASTM F1473, the "PENT" test, has been particularly well researched as a method to assess the resistance of a PE compound to slow crack growth, the second stage of environmental stress cracking. Materials that do not fail under ASTM D1693 after thousands of hours are more effectively characterized under ASTM F1473.

Under ASTM F1473, specimens are prepared from compression-molded plaques of PE resin or taken from pipe. Extremely sharp razor notches are cut across the specimen to a specified depth of the specimen thickness. The specimen is placed in a constant temperature air oven at 80°C, and a constant tensile stress of 2.4 MPa (348 psi) is applied to the unnotched area. The time to specimen breakage is measured. It should be noted that elevated temperature air is known to be an aggressive, oxidizing environment for PE, especially under applied stress.

An empirical study of PE pressure piping materials compared ASTM F1473 performance to service life and concluded that a failure time of 12 hours under ASTM F1473 compared to a service history of 50 years[11]. However, a minimum ASTM D3350 SCG cell classification value of 6, a minimum average failure time of 100 hours per ASTM F1473, is recommended for PE water pipes. This performance level provides a considerable margin against the potential for environmental stress-cracking failure in the field.

The final mode of pipe failure is Stage III or brittle-oxidative, which is the result of oxidative degradation of the polymer's material's properties. This type of failure is typically obtained under conditions of extreme laboratory testing. As a further precaution against Type III failure, the HDPE pipe industry has investigated the resistance of these products to failure under conditions of flowing potable water service. ASTM F2263[12] provides that pipe specimens are subjected to flowing water at specific conditions of temperature, pH, and chlorine content. These extreme test conditions are used to further improve the capabilities of HDPE piping systems and their ultimate resistance to environmental stress cracking in potable water applications.

In summary, PE pressure piping materials used in AWWA pressure piping are exceptionally resistant to environmental stress crack initiation and to slow crack growth if a crack does initiate. ANSI/AWWA C906 requires PE materials that are highly resistant to environmental stress cracking and establishes product tests to ensure against crack initiation sites in the pipe ID. As such, PE pipe produced and labeled with the ANSI/AWWA C901/C906 designation indicates that the product has been manufactured from a material that has been tested and found to meet or surpass the requirements for resistance to slow crack growth in either of these standards. For further information regarding the evolution of resistance to slow crack growth in PE pipe, see the references at the end of this chapter[13].

ENVIRONMENTAL CONSIDERATIONS

Weathering

Over time, ultraviolet (UV) radiation and oxygen may induce degradation in plastics that can adversely affect their physical and mechanical properties. To prevent this, various types of stabilizers and additives are compounded into a polymer to give it protection from these phenomena.

The primary UV stabilizer used in the PE pipe industry is carbon black, which is the most effective additive capable of inhibiting UV induced reactions. Carbon black is extremely stable when exposed to the outdoor elements for long periods of time and is relatively inexpensive compared to some of the more exotic colorant systems. The result is a piping system of uniform color that does not chalk, scale, or generate dust in response to extended periods of outdoor exposure.

PE pipe is generally formulated to resist ultraviolet (UV) degradation. Exposure to UV radiation leads to the formation of free radicals within the polymer matrix. These free radicals are then available to react with other molecules within the polymer, and the result can be a significant reduction in physical properties. The carbon black present in PE pipe acts as a primary UV absorber thus precluding the formation of free radicals. In this way, UV degradation is prevented, and the physical properties of the polymer are retained even after substantially long periods of exposure to the elements. Studies conducted by Bell Laboratories on the stability of carbon black containing PE used in wire and cable application have shown that these materials can sustain exposure to the elements over periods of 30 years plus with no appreciable change in the performance characteristics of the polymer[14].

While carbon black is a very effective UV screen that provides maximum UV protection, the degree of protection it imparts may not be required for buried pipe applications. Generally, UV protection is only required for relatively short periods of time while the pipe is exposed to sunlight such as during storage or while in transit or in the process of handling during installation. As a result, alternate UV stabilization systems have been developed that have proven very effective and permit the use of colored, nonblack PE pipe. The reader is referred to the pipe manufacturer for information regarding the availability of these nonblack products.

Stabilization

Prolonged exposure to excessive heat can also initiate the generation of free radicals in a polymer. A chemical stabilizer system is typically added to the PE to prevent the generation of these free radicals. Generally, these stabilization systems are produced from a combination of carbon black, FDA-approved antioxidants and heat stabilizers, or in the case of nonblack pipe, a series of FDA-approved heat stabilizers and antioxidants. These stabilization systems are designed and selected with the intention of providing long-term protection of the PE polymer from oxidation and thermal degradation. As noted, these additives are generally FDA approved and their suitability of use in potable water applications is determined in accordance with third party standards developed by a consortium that includes NSF International, American Water Works Association, Awwa Research Foundation, and other groups. The effectiveness of the stabilization system may be evaluated using differential scanning calorimetry (DSC) and/or the carbonyl index test. The DSC test measures the induction time to the onset of degradation and the temperature at which degradation begins. The carbonyl index test measures the degree of oxidative degradation by measuring the type and amount of carbonyl functional groups created on the surface of the polymer as a result of excessive exposure to heat or UV radiation.

PE pipe produced in accordance with ANSI/AWWA C906 must meet the requirements of ASTM D3350[15]. This industry recognized standard requires that the induction temperature for the onset of degradation must exceed 220°C.

Biological

Biological attack may be described as the degradation of the piping material caused by the action of organisms such as bacteria, fungi, insects, or rodents. PE has no nutritional value. It is considered inert in that it will neither support nor deter the growth or propagation of micro- or macro-organisms.

Numerous studies have been conducted over the years relative to the biological implications of PE pipe. These studies have revealed that insects or microorganisms pose no threat of damage or degradation to PE pipe. Some indication of rodent damage has been reported but most of this was related to placement of small diameter tubing in rodent infested areas. The resulting damage was attributed to the need for the rodent to maintain their teeth in good condition and the damage associated with gnawing on the profile was felt to be no greater with PE than with any other piping materials installed in these areas. Additional information is available from PPI Technical Report TR-11[16].

LONG-TERM PROPERTIES

Long-Term Hydrostatic Strength

The pressure capability of PE pipe is based on an extrapolation of stress-rupture data over time. The extrapolation method predominantly used in North America is defined in ASTM D2837[17]. Using this protocol, stress-rupture data at a specific temperature is gathered over a 10,000 hour period. If the data meets certain distribution criteria, the data is extrapolated to 100,000 hours. The stress intercept that is extrapolated at 100,000 hours is referred to as the *long-term hydrostatic strength* (LTHS) of the material being evaluated. The LTHS will fall into one of a series of preferred stress ranges defined in ASTM D2837. The category into which the LTHS falls is referred to as the *hydrostatic design basis* (HDB) of the material. It is this value that is used to determine the pressure capability of a pipe under specified service conditions. The designer is referred to ASTM D2837 for a complete listing of the categories and stress ranges that are used to establish the HDB for thermoplastic materials. The HDB is used to determine the pressure capability of a specific pipe profile or DR under certain conditions of stress. This pressure rating methodology is discussed in detail in Chapter 4.

The HDB for a material can be obtained at any of a variety of service temperatures. In fact, it is common practice to evaluate PE at the standard laboratory temperature of 73°F (23°C) and an elevated temperature of 140°F (60°C). Through a statistical analysis of the nature of both of the curves, information regarding the performance of the material under other service temperatures can be determined.

Information such as this is used by the Hydrostatic Stress Board (HSB) of the Plastics Pipe Institute (PPI) to issue recommendations for the HDB of thermoplastics materials that will be used to produce plastic pipe. These recommendations are reviewed and published periodically in PPI's TR-4[18].

Fracture Mechanics

Fracture mechanics refers to the study of crack growth originating from flaws that may exist within a material or structure. Flaws may be the result of inhomogenieties within a material, manufacturing inconsistencies, gouges, and scrapes that result from the handling or mishandling of the finished product or any other number of sources.

These flaws, whether microscopic or macroscopic in nature, act to intensify any nominal stress applied within the localized region. At some point, this intensified stress at the flaw will exceed the strength of the material and a small crack may develop. An initiated crack may subsequently grow and lead to failure of the part or component. Modern PE materials formulated specifically for pressure pipe applications are designed to resist the initiation of this slow crack growth phenomena even when subjected to millions of cycles of pressure transients.

The fracture resistance of a given structure or material will depend on the level of stress applied to it, the presence and size of any flaws in it, and the inherent resistance of the material to crack initiation and growth. Extensive research conducted on gas pipe indicates that modern PE resins designed for pressure piping applications are extremely resistant to slow crack growth[19]. The requirements of ANSI/AWWA C906 ensure that water pipe produced in accordance with this standard will demonstrate comparable levels of resistance to slow crack growth provided that the pipe system is designed, installed, and operated in accordance with the guidelines stated in subsequent chapters of this manual.

Fatigue

Each time a PE pipe is pressurized or subjected to hydraulic transients, its circumference expands and unrestrained length decreases in an elastic manner. For applications where the pressure is constant and below the pipe pressure rating, this small amount of expansion (strain) is not important and is not considered a design variable. However, strain does become important when the pipe undergoes higher, cyclic pressurization. There is a maximum critical strain limit, which once exceeded, permanently changes the characteristics of the pipe.

At higher strain levels, microcracks can develop within the PE matrix. Repeated straining that approaches the critical strain limit of the material can cause growth of the microcracks that may eventually propagate into a failure.

For modern PEs used in piping applications, the critical strain limit has been established to be 6 to 7 percent depending on the exact nature of the polymer. The typical PE pressure pipe undergoes a strain of 0.5 to 1.0 percent when placed in service, that is, a safety factor of at least 6 to 1. This is well below the critical strain limit for modern

PE pipe resins. Even cyclic pressure surges of up to 100 percent of the operating pressure of the PE water pipe system do not exceed the critical strain limit for these highly ductile materials.

To this end, Bowman and Marshall (et al.) have conducted extensive research on the fatigue resistance of modern PE pipe compounds[20, 21]. Based on his research of combined creep and surge regimes at 80°C, Bowman concluded that butt-fused PE piping systems provide years of uncompromised service exceeding millions of surge cycles even under conditions of sustained pressurization. Marshall and his colleagues determined that today's tough PE pipe formulations can withstand sustained periods of high frequency surging (ranging from 1 to 50 cycles per hour) at magnitudes of up to 200 percent of the pipe's static pressure rating with no indication of fatigue and no reduction in long-term serviceability when properly installed. Research results such as these serve as the basis for the surge allowances stipulated in ANSI/AWWA C906.

INDUSTRY STANDARDS

Industry standards exist to establish the minimum level of performance for PE piping based on the physical properties resulting from the combined effect of the three fundamental polymer properties: density, molecular weight, and molecular weight distribution. Primary among these is ASTM D3350 and the various additional standards included by reference within ASTM D3350.

ASTM D3350 is a comprehensive classification standard that delineates seven key properties associated with piping performance. Ranges of performance for each of these properties are defined within this standard as well. The result is a matrix of piping related material properties defined by classification cells, which can be utilized to identify the particular PE compound used to manufacture pipe. Six properties and their respective cell limits are reproduced from ASTM D3350 in Table 1-3.

The seventh property, color and UV stabilizer, is identified by a letter, which follows the six cell classification numbers described in Table 1-3. The code letters for color and UV stabilizer are

A—for natural
B—for colored
C—for black with minimum 2 percent carbon black
D—for natural with UV stabilizers
E—for colored with UV stabilizers

PE pipe compounds are typically black with a minimum of 2 percent carbon black, C, or colored with UV stabilizers, E.

By utilizing ASTM D3350, the pipe designer can reference a specific combination of six numbers and one letter to establish a minimum level of performance based on the properties referenced in the standard. Pipe produced in accordance with ANSI/AWWA C906 must be manufactured from PE material, which meets one of three cell classifications[21] as defined by ASTM D3350. As noted in ANSI/AWWA C906, higher cell classes for some of the properties are allowed, but those for density and HDB are not. Considering the case for one of these cell classifications, 345464C, Table 1-4 presents an explanation of the minimum level of performance to be met.

By specifying a pipe produced from a feedstock with a cell classification of 345464C, as shown in Table 1-3, the designer has established the minimum level of performance for the polymer from which the pipe is produced. For example, in Table 1-4, a base resin density of 0.941 to 0.955 g/cc per ASTM D1505 is required. Base resin density is one of the key molecular parameters for PE in piping applications.

Further, the designer has specified a polymer with a melt index of less than 0.15 gr/10 min in accordance with ASTM D1238. Melt index is a relative indication of the

Table 1-3 ASTM D3350 cell classification limits

Property	Test Method	0	1	2	3	4	5	6
1. Density, g/cm^3	D1505	—	0.910–0.925	0.926–0.940	0.941–0.947	0.948–0.955	>0.955	—
2. Melt index, g/10 min	D1238	—	>1.0	1.0 to 0.4	<0.4 to 0.15	<0.15	A	B
3. Flexural modulus, Mpa (psi)	D790	—	<138 (<20,000)	138<276 (20,000 to <40,000)	276<552 (40,000 to <80,000)	552<758 (80,000 to <110,000)	758<1,103 (110,000 to <160,000)	>1,103 (>160,000)
4. Tensile Strength at Yield, Mpa (psi)	D638	—	<15 (<2,200)	15<18 (2,200 to <2,600)	18<21 (2,600 to <3,000)	21<24 (3,000 to <3,500)	24<28 (3,500 to <4,000)	>28 (>4,000)
5. Slow crack growth resistance								
I. ESCR	D1693							
a. Test condition			A	B	C	C	—	—
b. Test duration, hr		—	48	24	192	600		
c. Failure, max., %			50	50	20	20		
II. PENT (hr) Molded plaque, 80°C, 2.4 Mpa, Notch depth per F14732, Table I	F1473		0.1	1	3	10	30	100
6. HDB, MPa (psi) @ 23°C	D2837	—	5.52 (800)	6.89 (1,000)	8.62 (1,250)	11.03 (1,600)	—	—

Table 1-4 Example of D3350 cell class specification

Property	ASTM Method	Cell Class Designation
Density	D1505	3
Melt index	D1238	4
Flexural modulus	D790	5
Tensile strength	D638	4
Slow crack growth resistance	F1473	6
HDB	D2837	4
Color and UV stabilizer code	D3350	C

molecular weight of the polymer, and the two are inversely related. That is, a lower melt index suggests a higher molecular weight. A higher molecular weight relates to an increase in certain physical properties as shown in Table 1-1.

The relative stiffness of the piping material as reflected in the flexural modulus is specified to be between 110,000 and 160,000 psi per ASTM D790 by the designation of a 5 in the third character position of the D3350 cell class designation.

By using the 345464C cell class designation, the short-term tensile yield strength of the polymer used to manufacture the pipe has been specified to be between 3,000 and

3,500 psi as shown in Table 1-3. Yield strength is determined in accordance with ASTM D638.

The ability of the pipe to resist slow crack growth is an important engineering design consideration. ASTM D3350 allows for the determination of slow crack resistance by either the D1693 (bent strip) method or the F1473 (PENT) method. Piping products produced from resins that maintain a slow crack growth cell classification of 4 have been successfully used for decades in potable water applications. However, recent advances in polymer technology provide for modern PE pipe resins that maintain cell classes of 5 and 6, which require testing in accordance with ASTM F1473, thus assuring the end user of even higher levels of technical performance as it relates to the slow crack growth resistance of modern PE pipe. A specification for modern PE piping systems that designates a material requirement of 345464C, D, or E identifies the highest level of resistance to slow crack growth.

The final numerical designation relates to the hydrostatic design basis of the material used to produce the pipe so specified. In this particular example, the designer has stipulated a material with a minimum hydrostatic design basis of 1,600 psi established in accordance with ASTM D2837. It is this value that is subsequently utilized to establish the pressure capability of the pipe based on the relationship between stress, wall thickness, and diameter.

The letter designation in the D3350 method of classification establishes a requirement for color and/or stabilization against the deleterious effects of UV light. In this particular case, the letter C refers to a material that must contain a minimum of 2 percent carbon black. The carbon black not only acts as a colorant but also as a primary UV stabilizer.

While ASTM D3350 provides for a significant number of combinations of cell classifications, not all combinations are commercially available. Additional information regarding the commercial availability of various cell class combinations may be obtained from the pipe producers and/or resin manufacturers. Additionally, ANSI/AWWA C901 and C906 refer to three specific minimum cell classifications, which may be used in potable water applications and, as such, the reader is referred to these standards for clarification on the cell class combinations, which are suitable for these installations.

PE is a thermoplastic material and may therefore be manufactured into a product, then ground into particles and remanufactured into another product. Once a product has been reduced to particles of an appropriate size, the material is called *rework* or *regrind.* Rework material for plastic pipe manufacturing is defined in ASTM F412.

ANSI/AWWA C901 and C906 allow for the use of rework material in the production of PE pipe with very specific limitations:

> "Clean rework materials derived from the manufacturer's own pipe or fitting product may be used by the same manufacturer for similar purposes provided that
>
> 1) The cell classification of the rework material is identical with the materials to which it will be added;
>
> 2) The rework material complies with all the applicable requirements of section 4.2 of this standard;
>
> 3) The finished products meet the requirements specified by the purchaser and comply with all requirements of this standard."

Using regrind or rework material does not adversely affect the resulting pipe or pipe fittings provided the commonly accepted material handling procedures are followed. These procedures ensure the cleanliness and segregation of the materials until they are incorporated in the product manufacturing process.

The requirement in ANSI/AWWA C901 and C906 for "clean rework materials derived from a manufacturer's own pipe or fitting" prevents the use of PE material that has left the control of the original manufacturer, i.e., pipe products that were in service and were replaced.

CONCLUSION

The information contained in this chapter is provided to assist the reader in understanding some of the fundamental properties of PE. A basic understanding of these properties will assist the pipe designer in the effective use of these materials and serve to maximize the utility of the service into which they are ultimately placed. For further information concerning the engineering properties of PE piping, the reader is referred to a variety of sources including pipe manufacturer's literature, additional publications of the PPI, and the references at the end of this chapter.

REFERENCES

1. ASTM D1238, *Standard Test Method for Flow Rates of Thermoplastics by Extrusion Plastometer,* ASTM International, West Conshohocken, PA.
2. ASTM D638, *Standard Test Method for Tensile Properties of Plastics,* ASTM International, West Conshohocken, PA.
3. ASTM D695, *Standard Test Method for Compressive Properties of Rigid Plastics,* ASTM International, West Conshohocken, PA.
4. ASTM D790, *Standard Test Methods for Flexural Properties or Unreinforced and Reinforced Plastics and Electrical Insulating Materials,* ASTM International, West Conshohocken, PA.
5. ASTM D256, *Standard Test Method for Determining the Pendulum Impact Resistance of Notched Specimens of Plastics,* ASTM International, West Conshohocken, PA.
6. ANSI/AWWA C906, *AWWA Standard for Polyethylene (PE) Pressure Pipe and Fittings, 4 In. (100 mm) Through 63 In. (1,575 mm), for Water Distribution and Transmission,* American Water Works Association, Denver, CO.
7. "Engineering Properties of Polyethylene," *The Handbook of Polyethylene Pipe,* Plastics Pipe Institute, Washington DC.
8. PPI TR-19, *Thermoplastics Piping for the Transport of Chemicals,* Plastics Pipe Institute, Washington DC.
9. ASTM D1693, *Standard Test Method of Environmental Stress-Cracking of Ethylene Plastics,* ASTM International, West Conshohocken, PA.
10. ASTM F1473, *Standard Test Method for Notch Tensile Test to Measure the Resistance to Slow Crack Growth of Polyethylene Pipes and Resins,* ASTM International, West Conshohocken, PA.
11. Brown, N. and X. Lu, *PENT Quality Control Test for PE Gas Pipes and Resins,* Department of Materials Science and Engineering, University of Pennsylvania, Philadelphia, PA 19104-6272.
12. ASTM F2263, *Standard Test Method for Evaluating the Oxidative Resistance of Polyethylene (PE) Pipe to Chlorinated Water,* ASTM International, West Conshohocken, PA.
13. Sandstrum, S.D., "Polyethylene Gas Pipe... Here Today, Here Tomorrow," Proceedings of the 17th International Plastic Fuel Gas Pipe Symposium, American Gas Association, Washington DC, 2002.
14. Gilroy, H.M., 1985. "Polyolefin Longevity for Telephone Service," ANTEC Proceedings, Society of Plastic Engineers, Brookfield, CT.
15. ASTM D3350, *Standard Specification for Polyethylene Plastics Pipe and Fittings Materials,* ASTM International, West Conshohocken, PA.
16. PPI TR-11, *Resistance of Thermoplastics Piping Materials to Micro- and Macro-Biological Attack,* Plastics Pipe Institute, Washington DC.
17. ASTM D2837, *Standard Test Method for Obtaining Hydrostatic Design Basis for Thermoplastic Pipe Materials,* ASTM International, West Conshohocken, PA.

18. PPI TR-4, *PPI Listing of Hydrostatic Design Basis (HDB), Strength Design Basis (SDB), Pressure Design Basis (PDB) and Minimum Required Strengths (MRS) for Thermoplastics Piping Materials for Pipe,* Plastics Pipe Institute, Washington DC.
19. Lustiger, A. and N. Ishikawa, "Measuring Relative Tie Molecule Concentration in Polyethylene," Proceedings of the 11th Plastic Fuel Gas Pipe Symposium, American Gas Association, Washington DC, 1989.
20. Bowman, J.A., "The Fatigue Response of Polyvinyl Chloride and Polyethylene Pipe Systems," *Buried Plastics Pipe Technology, ASTM STP1093,* American Society for Testing and Materials, Philadelphia, 1990.
21. Marshall, G.P., S. Brogden, M.A. Shepherd, "Evaluation of the Surge and Fatigue Resistance of PVC and PE Pipeline Materials for Use in the UK Water Industry," Proceedings of Plastics Pipes X, Goteborg, Sweden.

This page intentionally blank.

Chapter 2

Manufacturing, Testing, and Inspection

INTRODUCTION

As stated in Chapter 1, PE belongs to a group of polymers called thermoplastics. Thermoplastics first soften and then melt with sufficient heating and reharden when cooled. The thermoplastic behavior of PE allows it to be readily shaped by processes such as extrusion, molding, or thermoforming into pipe, fittings, and parts for other applications. Parts processed from PE are essentially isotropic in their properties. Regardless of the orientation of the part or stress applied, properties such as tensile strength and resistance to slow crack growth are essentially the same. This makes it excellent for use in molded parts with complex shapes. In some special molding or production circumstances, PE's isotropy can be reduced, but this reduction can be compensated for and even used to improve material properties for particular applications.

PE is quite viscous in the molten state and can be easily fused to itself. This feature allows PE pipe and fittings to be fabricated by the thermal fusion of separate components. Thermal fusion is also the basis for the butt-, socket-, and electro-fusion processes, which are used for the joining of PE fittings to pipe.

To protect PE against thermal and oxidative degradation during fabrication and ensure a long shelf and service life, PE is always compounded with a small amount of additives. Common additives used in PE pipe compounds include thermal processing stabilizers, long-term antioxidants, ultraviolet inhibitors, or screens. Other additives such as pigments can also be added to modify the appearance or physical properties of the polymer.

PE pipe resins are generally available to the pipe and fittings manufacturer in the form of natural pellets, whereby all required ingredients except for pigment are premixed and predispersed into the base polymer. These pelletized compounds are commonly delivered to the piping manufacturer by hopper trucks or tank cars, and off-loaded into silos and transferred to the fabricating equipment by pneumatic conveying systems.

The natural pellets are then mixed with an approved color concentrate at the pipe or fitting production site to result in an appropriately colored finished product.

PIPE MANUFACTURE

PE pressure pipes are manufactured by the extrusion process, whereby molten polymer material is continuously forced through an annular die by a turning screw. As the screw rotates, the flights on the screw continually convey material closer and closer to the die. The pipe is shaped as it exits the die and into the downstream equipment.

Although simple in concept, the extrusion process used in the manufacture of plastic pipe employs a series of fairly sophisticated equipment that reliably and continuously performs the functions of feeding, melting, and mixing the material, delivering it to the die at appropriate and consistent conditions, accurately shaping and sizing the pipe, and cooling and cutting or coiling the product to precise lengths. As illustrated in Figure 2-1, this series of equipment, commonly referred to as an extrusion line, generally consists of the extruder, the pipe die, sizing and cooling tanks, pipe puller, saw, and take-off equipment.

The extruder (Figure 2-2) is used to mix and heat the material to the prescribed temperature and to force the molten polymer through the pipe extrusion die. The source of heat to initially melt the polymer during start-up is usually a series of heater bands wrapped around the extruder barrel. However, once the extruder is operating at steady-state, almost all of the heat needed to melt the PE pellet feed comes from friction caused by the shearing of the PE material by the revolving screw while it is mixing and compressing the plastic. Cooling fans or water hook-ups are also usually attached to the barrel of the extruder in the event cooling is needed.

Throughout the length of the screw, the temperature of the plastic is closely monitored and controlled. The temperature settings depend a great deal on throughput rates, screw and die geometry, and the limits of the material being extruded.

The pressure of the molten PE at the end of the extruder barrel before it enters the die can range from 1,000 to 5,000 psi. The melt pressure is dependent on the rheology of the PE being used and the geometry and speed of the extrusion line.

The melt then passes through a screen pack that consists of one or more fine mesh screens before entering the die. The screen pack prevents foreign contaminants from entering the pipe wall, and it also generates some additional back pressure in the polymer melt further enhancing the quality of mixing.

The die shapes the extrusion melt into a cylinder of slightly greater size than the final pipe product. When the PE melt exits the die, it is still molten, but the high viscosity of the material allows it to retain its shape. The final dimensions of the solid pipe are set by the sizing and cooling operation downstream of the die. Sizing is accomplished by pulling the hot PE cylinder through an accurately bored sizing sleeve that fixes the pipe's outside diameter. The pipe's outer surface is tightly held against this sleeve by applying either vacuum on the outside of the pipe or internal pressure on the inside of the pipe. While the pipe is being sized, it is also being cooled with water to set the outside diameter. The pipe must be cooled even after the outside diameter has been set to remove heat stored in the pipe wall. To accomplish this, one or more cooling tanks are set up downstream of the sizing sleeve/tank, and the pipe is pulled through each.

A puller placed after the cooling tanks provides the necessary force to pull the pipe through the entire cooling operation. The pulling rate in relation to the extrusion rate is very carefully set and maintained for accurate control of pipe wall thickness. The pulling rate is usually slightly faster than the extrusion rate to draw down the wall.

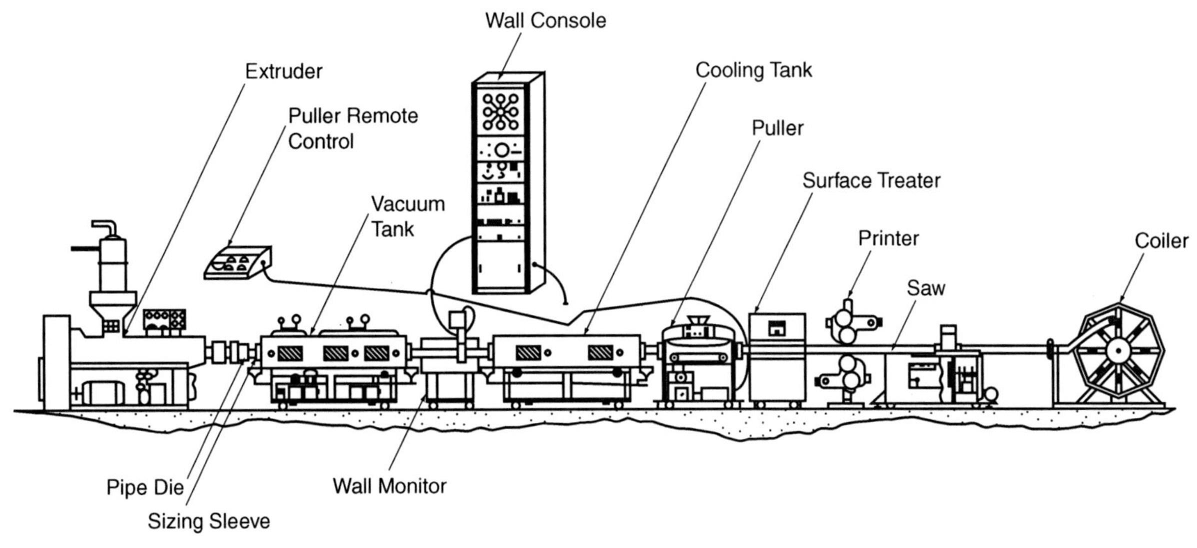

Courtesy of the Plastics Pipe Institute

Figure 2-1 Typical extrusion line

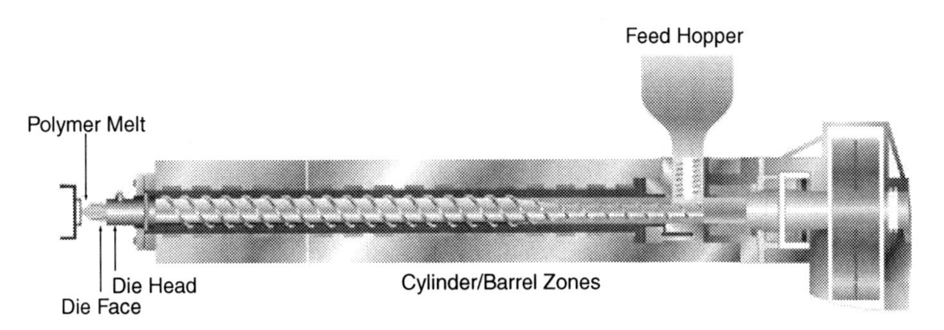

Courtesy of BP Solvay Polyethylene N.A.

Figure 2-2 Typical pipe extruder

This practice promotes consistency in wall thickness and some orientation to the polymer melt.

A printer for marking the pipe is usually located either immediately before or after the puller. In accordance with AWWA standards, the marking must be placed at frequent intervals and identify the pipe material, nominal pipe size, pipe class or dimension ratio (DR), the manufacturer, and a manufacturing code that provides specific information such as manufacturing date, location, or other data to assist in the traceability of the product.

Beyond the printer, the pipe is cut to the correct length and dropped into a storage or bundling cart. Small diameter pipe such as 6-in. nominal diameter and smaller is available coiled for handling, shipping, and installation convenience.

FITTINGS MANUFACTURE

Plastic pipe fittings, because of their complex geometry, are generally manufactured by the injection molding process. This process is similar to extrusion in that the major operational steps are the same: mixing and melting of the material, then shaping and

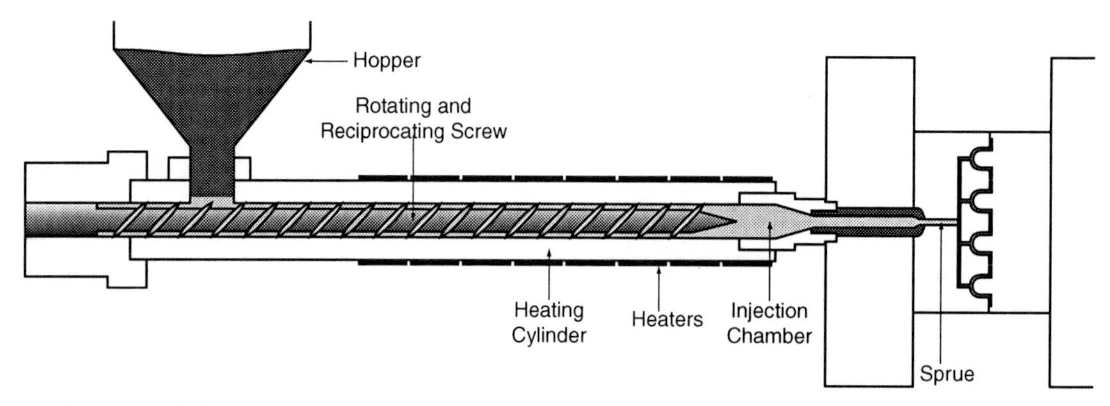

Courtesy of BP Solvay Polyethylene N.A.

Figure 2-3 Typical injection molding machine

Figure 2-4 Prefabricated 90° elbow being attached in field

sizing, and finally cooling of the part. While these steps are continuous in extrusion, they are performed cyclically in injection molding (Figure 2-3).

First, incoming resin in the form of pellets is melted by a combination extruder–injection screw. The screw turns to thoroughly melt and mix the pellets while conveying the material to the tip of the screw. The screw plunges forward to force the molten PE through the die and into a closed, heated mold. In the next stage, the full mold is cooled by circulating cold water while the screw maintains pressure on the mold. Once the part is thoroughly cooled and solidified, the mold is opened, and the finished part is removed.

Injection molded fittings include tees, 45° and 90° elbows, reducers, couplings, caps, flange adapters and stub ends, branch and service saddles, and self-tapping saddle tees. These molded fittings are designed to be applied on the pipe by means of socket-, butt-sidewall or the electro-fusion process (Figure 2-4). The details of the heat fusion process are discussed in Chapter 6.

Fittings can also be manually fabricated. Fabricated fittings are constructed by heat fusing sections of pipe, machined blocks, or molded fittings to produce the desired configuration. Because of tests required under AWWA standards, fittings such as elbows cannot be fabricated in the field.

Fabricated fittings intended for full pressure service must be designed with additional material in areas subject to high stress. The common commercial practice is to increase wall thickness in high-stress areas by using heavier wall pipe sections. This is similar to molded fittings that are molded with heavier body wall thickness.

Fittings can also be fabricated using the thermoforming process by heating a section of pipe and using a forming tool to reshape the heated area. Examples are sweep elbows, swaged reducers, and forged stub ends. The area to be shaped is immersed in a hot liquid bath and heated to make it pliable. It is removed from the heating bath and reshaped in the forming tool. The new shape is held until the part has cooled.

Very large diameter fabricated fittings (> 12 in. in diameter) require special handling during shipping, unloading, and installation. Precautions should be taken to prevent bending moments that could break the fitting during these periods. These fittings are sometimes wrapped with a reinforcement material, such as fiberglass, for protection. Reinforcing material is used only for mechanical protection, not to increase the design pressure rating of the fitting.

TESTING AND INSPECTION

The purpose of testing and inspecting raw materials and finished products in the manufacturing of PE piping can be grouped into three categories: 1) to ensure that materials and processes used are qualified for the intended product and service; 2) for quality control of incoming materials and of manufacturing processes; and 3) for assurance that the end product, after manufacture, is in compliance with the referenced standards. To this end, many pipe producers are certified in accordance with the requirements of ISO 9000[1]. This internationally recognized standard on quality policies and procedures details specific criteria which the manufacturer must verify compliance with customer and industry specifications.

Qualification testing documents the long-term properties of the PE resin and the process conditions for yielding a product that is suitable for the intended service and is in compliance with the applicable standards. Quality control (QC) testing is conducted during manufacture and provides timely feedback used for maintaining the process and product performance at a high level. Quality assurance testing, which often takes longer and is more complex to perform than is practical for QC purposes, is conducted to give greater and final assurance of product quality and compliance.

Qualification Testing

There are qualification requirements imposed on both the material and end product. PE used for the manufacture of pipe and tubing in accordance with AWWA standards must demonstrate the following qualifications:

Health effects evaluation. All materials for pipe and fittings must comply with ANSI/NSF Standard 61[2]. This document, which was developed by a consortium that included the American Water Works Association and the Awwa Research Foundation, establishes minimum requirements for the control of potentially adverse human health effects from products that contact drinking water. To qualify under this standard, testing must be conducted to verify that substances cannot be extracted by water in quantities that can reasonably be determined to result in toxic, carcinogenic, eratogenic, or mutagenic effects in humans who may be consuming water that comes in contact with products made from the subject material.

Long-term hydrostatic strength. Materials must be tested to establish that their minimum long-term hydrostatic strength is in compliance with the requirements of the applicable AWWA product standard. ANSI/AWWA C901[3] and C906[4] standards allow materials listed in PPI TR-4[5]. To obtain a listing in this document, resin manufacturers are required to submit stress-rupture data developed in accordance with PPI TR-3[6]. Based on such data, PPI assigns the subject material a hydrostatic design basis (HDB) and lists the material's trade name and corresponding HDBs in PPI TR-4.

Material cell classification. ANSI/AWWA C901 and C906 also require that PE materials have a cell classification per ASTM D3350[7]. This specification uses a cell format whereby materials are classified according to certain physical properties (i.e., density, flexural modulus, tensile strength, etc.). AWWA standards also require a minimum level of antioxidants and ultraviolet (UV) radiation protection.

Additional qualification requirements for PE pipe and fittings are detailed in ANSI/AWWA C901 and C906 along with their referenced ASTM standards. ISO 9000 certification of the manufacturer producing parts to these standards requires that policies, procedures, and compliance documentation be maintained to verify that the requirements established within these standards are met.

Quality Control Testing

QC testing is conducted both on the incoming plastic piping materials and the manufactured pipe and fittings. The intent of the material QC tests is to determine that the materials comply with AWWA requirements and that the material properties also match the incoming certificate of analysis from the resin manufacturer. The intent of the latter is to ensure that the incoming material type is qualified for pipe and fitting manufacture. Typical material QC tests include density per ASTM D1505[8] and melt index per ASTM D1238[9].

PE pipe manufacturing is a continuous process. Therefore, the pipe producer generally monitors those process control variables that are determined as quality critical. Detailed records of these production variables are kept as a means to monitor the pipe production process thereby ensuring the quality of the pipe. Additionally, QC tests are conducted on the pipe during production on a regular basis to further ensure the quality of the finished product. These tests include but are not limited to the following:

- Dimensions. Pipe diameter, wall thickness, ovality, and length are periodically measured in accordance with ASTM D2122[10] to ensure compliance. All fittings also are periodically checked for proper dimensions and tolerances.
- Workmanship. All pipe, tubing, and fittings are routinely inspected to ensure that they are homogenous throughout—free of visible cracks, holes, foreign inclusions, blisters, dents, or other injurious defects. They are also checked to ensure that they are as uniform as commercially practical in color opacity, density, and physical properties.
- Pressure tests. Pipes are subject to one of the following pressure tests:
  a. The quick-burst test, conducted in accordance with ASTM D1599[11]. This test is one measure of production quality. The failure should be ductile and above a certain pressure threshold depending on material and pipe dimensions.
  b. A pipe is exposed to a pressure of four times its pressure class for a period of five seconds. When the pipe or part being tested survives, it is still in compliance with the dimensional requirements of the standard.
  c. The ring tensile test is conducted on a ring cut from the extruded pipe. It is designed to establish that the tensile properties of the pipe (i.e., the tensile strength and elongation at failure) are relatively consistent with

the raw material properties. As in the case of the quick-burst test, lack of ductility signals a problem.

- Bend-back test. The bend-back test evaluates the inside pipe surface for brittleness. The evaluation is made by careful visual inspection of the inside surface of an arc section of the pipe that has been highly strained by bending against the curvature of the pipe wall. The presence of surface cracking or crazing signals that this surface may have become thermally degraded by excessive extrusion temperatures.
- Marking. All pipe and fittings are frequently checked to ensure that they are clearly and appropriately marked in accordance with the applicable AWWA standard.

Quality Assurance Testing

Quality assurance (QA) testing is performed on a statistical basis at the completion of the manufacturing process for further assurance that the product satisfies all applicable standards. Typical QA testing includes the following:

- Sustained-pressure test. In this test, pipe specimens must survive without failure a prescribed pressure applied for not less than 1,000 hr at 23°C.
- Elevated-temperature sustained-pressure test. In this test, the average failure time of three pipes and the failure time of two of the three pipes must exceed certain minimum requirements when tested at specified pressures and 80°C. See ANSI/AWWA C901 or C906 for additional information.
- Thermal stability test. In this test, a sample cut from extruded pipe is evaluated against the minimum stabilizer protection requirements to ensure that the extrusion process has not used up antioxidants and stabilizers that have been added to the raw materials. This is important for ensuring the long-term performance of the piping product.

CONCLUSION

This chapter provides information regarding the qualification, manufacturing, inspection, and testing systems used for the production of PE water pipe and fittings. A fundamental understanding of the production process and the quality assurance and qualification testing inherent to PE pipe should allow the designer or installer of these products to use them with confidence for their intended purpose in potable water applications. For additional information on these systems, please consult the piping or fittings manufacturer or the ASTM and AWWA standards referenced in the text.

REFERENCES

1. ISO 9000, *Quality Management Systems – Fundamentals and Vocabulary,* International Organization for Standardization (ISO), Geneva, Switzerland.
2. NSF/ANSI Standard 61, *Drinking Water System Components – Health Effects,* NSF International, Ann Arbor, MI.
3. ANSI/AWWA C901, *Polyethylene (PE) Pressure Pipe and Tubing, ½ In. (13 mm) Through 3 In. (76 mm), for Water Service,* American Water Works Association, Denver, CO.
4. ANSI/AWWA C906, *Polyethylene (PE) Pressure Pipe and Fittings, 4 In. (100 mm) Through 63 In. (1,575 mm), for Water Distribution and Transmission,* American Water Works Association, Denver, CO.
5. PPI TR-4, *PPI Listing of Hydrostatic Design Basis (HDB), Strength Design Basis (SDB), Pressure Design Basis (PDB)*

and Minimum Required Strength (MRS) Ratings for Thermoplastic Piping Materials or Pipe, Plastics Pipe Institute, Washington, DC.

6. PPI TR-3, *Policies and Procedures for Developing Hydrostatic Design Basis (HDB), Pressure Design Basis (PDB), Strength Design Basis (SDB), and Minimum Required Strengths (MRS) Ratings for Thermoplastic Piping Materials or Pipe,* Plastics Pipe Institute, Washington, DC.
7. ASTM D3350, *Standard Specification for Polyethylene Plastics Pipe and Fittings Materials,* ASTM International, West Conshohocken, PA.
8. ASTM D1505, *Standard Test Method for Density of Plastics by the Density-Gradient Technique,* ASTM International, West Conshohocken, PA.
9. ASTM D1238, *Standard Test Method for Flow Rates of Thermoplastics by Extrusion Plastometer,* ASTM International, West Conshohocken, PA.
10. ASTM D2122, *Standard Test Method for Determining Dimensions of Thermoplastic Pipe and Fittings,* ASTM International, West Conshohocken, PA.
11. ASTM D1599, *Standard Test Method for Resistance To Short-Time Hydraulic Pressure of Plastic Pipe, Tubing, and Fittings,* ASTM International, West Conshohocken, PA.

Chapter 3

Hydraulics of PE Pipe

INTRODUCTION

This chapter discusses the Darcy-Weisbach and Hazen-Williams flow equations for use with water flow in PE pipe. These equations depend, to some extent, on experimentally determined coefficients. The typical design friction coefficients for PE pipe are presented; however, the engineer may adjust the design friction factors based on experience and actual field conditions.

For the same change in flow velocity and for the same dimension ratio, PE pipes usually experience lower surge pressures than other types of pipe because of the lower elastic modulus of PE. Surge equations are discussed in Chapter 4.

DETERMINING THE FLOW DIAMETER OF A PE PIPE

Most PE pipe is made to controlled standard outside diameters and in either iron pipe sizes (IPS) or ductile iron pipe sizes (DIPS). When the outside pipe diameter is specified, the pipe's inside diameter is determined by the pipe's required wall thickness. The thicker the required wall, the smaller is the pipe's inside diameter.

The minimum wall thickness of a pipe is, in turn, determined by the pipe's specified dimension ratio (*DR*). By definition, the *DR* is the ratio of the specified average outside diameter of the pipe (D_o) to its minimum required wall thickness (t) or, $DR = D_o/t$. To ensure compliance with the minimum wall thickness requirements of applicable standards, the average wall thickness of PE pipe is slightly greater than the minimum. The typical tolerance for PE pipe wall thickness is plus 12 percent, so the accepted practice is to assume that the average wall (t_a) is 6 percent thicker than the specified minimum (t). According to this practice, the average inside pipe diameter (D_i) of pipe that is made to a standard outside diameter (D_o) is as follows:

$$D_i = D_o - 2t_a = D_o - 2(1.06t) \qquad \text{(Eq 3-1)}$$

Because t is determined by a pipe's *DR* (or *SDR*), then:

$$D_i = D_o - \frac{2.12D_o}{DR} \quad \text{(Eq 3-2)}$$

Table 3-1 and Table 3-2 give the dimension ratio, outside diameter, minimum wall thickness, average inside diameter, weight, and static pressure rating for IPS and DIPS pipe, respectively.

FRICTION HEAD LOSS

The frictional head loss that occurs in a moving, essentially incompressible fluid is given by the Darcy-Weisbach equation:

$$h_f = f \frac{v^2}{2g} \frac{L}{D} \quad \text{(Eq 3-3)}$$

Where:

h_f = frictional head loss, ft of liquid head

f = Darcy-Weisbach Friction Factor, dimensionless

v = average velocity of flowing fluid, ft/sec

g = acceleration due to gravity, 32.2 ft/sec^2

L = length of pipe, ft

D = pipe inside diameter, ft

The velocity of the flowing fluid (v) may be calculated from the rate of fluid flow and the pipe's average inside diameter. A commonly used formula is as follows:

$$v = 0.4085 \frac{Q}{D^2} \quad \text{(Eq 3-4)}$$

Where:

Q = flow rate, gal/min

DARCY-WEISBACH FRICTION FACTOR

Depending on the characteristics of the pipe surface, the nature and velocity of the flowing fluid, and the operating conditions, flow can occur in one of three regimes: laminar, transitional, or fully turbulent. Each of these three regimes is characterized by a different kind of response of frictional resistance to fluid flow that is presented by the inner pipe surface. These differences are illustrated by the Moody diagram (Figure 3-1) in which the Darcy-Weisbach factor of Eq 3-3 is plotted against the Reynolds number (Re), a parameter that is defined as follows:

$$\text{Re} = \frac{vD}{\gamma} \quad \text{(Eq 3-5)}$$

Where:

Re = Reynolds number, dimensionless

γ = kinematic viscosity of the flowing fluid, ft^2/sec (the ratio of dynamic viscosity to mass density)

D = average inside pipe diameter, ft

v = velocity of the flowing fluid, ft/sec

At low Reynolds numbers ($Re < 2{,}000$), flow is laminar and the friction factor, f, is not dependent on wall roughness. It is only a function of the Reynolds number:

$$f = \frac{64}{Re} \qquad \text{(Eq 3-6)}$$

Eq 3-6 plots as a straight line with a slope of −1 on the Moody diagram.

Between the laminar and transition regions lies a "critical zone." In this zone, which lies in the range of Reynolds numbers between approximately 2,000 and 4,000, the flow may be either laminar or turbulent depending on several factors that can include pipe roughness, changes in pipe cross-section and in direction of flow, and obstructions such as valves and fittings. The friction factor in this region is indeterminate and has lower and upper limits based on laminar flow and turbulent flow conditions.

At Reynolds numbers above approximately 4,000, flow conditions become more stable and definite friction factors can be established. The effect of both the Reynolds number and the roughness of the pipe wall on the friction factor in this region has been defined based on extensive experimental data. The results are incorporated into the Moody diagram. Colebrook developed the following empirically derived equation for predicting the friction factor in the transition region between smooth and fully turbulent flow:

$$\frac{1}{\sqrt{f}} = -2\log_{10}\left[\frac{\varepsilon}{3.7D} + \frac{2.51}{Re\sqrt{f}}\right] \qquad \text{(Eq 3-7)}$$

Where:

ε = absolute roughness of the pipe, ft

As noted on the Moody diagram (Figure 3-1), the quantity ε/D in the Colebrook equation is referred to as the relative roughness.

Beyond the transition zone is the zone of complete turbulence in which relative pipe roughness (ε/D) is essentially independent of the Reynolds number. It should also be noted from inspection of the Moody diagram that for pipes of very low relative roughness, so called "smooth" pipes, flow does not enter the zone of complete turbulence, except at very high Reynolds numbers.

The absolute roughness (ε) of PE pipe is 0.000005 ft. The surface characteristic of PE pipe is classified as "smooth" pipe and as such, it offers minimal resistance to the flow of fluids. Because PE piping is immune to corrosive attack by water and other substances, and because it is less likely to accumulate scaling or tuberculation, its original "smooth pipe" properties tend to be retained through the pipe's entire service life.

Table 3-1 PE 3408 polyethylene pipe iron pipe size (IPS) pipe data

Pipe weights are calculated in accordance with PPI TR-7 using a density of 0.955. Average inside diameter is calculated using nominal OD and minimum wall plus 6% for use in estimating fluid flows. When designing components to fit the pipe ID, refer to pipe dimensions and tolerances in applicable pipe specifications.

Pressure ratings are for water at 73.4°F. For other fluid and service temperature, ratings may differ.

Pressure Class		254 psi DR 7.3			200 psi DR 9.0			160 psi DR 11.0			128 psi DR 13.5			
IPS Pipe Size (in.)	Average D_o (in.)	Minimum Wall, t (in.)	Average D_i (in.)	Weight, w_P (lb/ft)	Minimum Wall, t (in.)	Average D_i (in.)	Weight, w_P (lb/ft)	Minimum Wall, t (in.)	Average D_i (in.)	Weight, w_P (lb/ft)	Minimum Wall, t (in.)	Average D_i (in.)	Weight, w_P (lb/ft)	IPS Pipe Size (in.)
1 1/4	1.660	0.227	1.179	0.44	0.184	1.270	0.37	0.151	1.340	0.31	0.123	1.399	0.26	1 1/4
1 1/2	1.900	0.260	1.349	0.58	0.211	1.453	0.49	0.173	1.533	0.41	0.141	1.601	0.34	1 1/2
2	2.375	0.325	1.686	0.91	0.264	1.815	0.76	0.216	1.917	0.64	0.176	2.002	0.53	2
3	3.500	0.479	2.485	1.98	0.389	2.675	1.66	0.318	2.826	1.39	0.259	2.951	1.15	3
4	4.500	0.616	3.194	3.27	0.500	3.440	2.74	0.409	3.633	2.29	0.333	3.794	1.90	4
5 3/8	5.375	0.736	3.815	4.66	0.597	4.109	3.90	0.489	4.338	3.27	0.398	4.531	2.72	5 3/8
5	5.563	0.762	3.948	5.00	0.618	4.253	4.18	0.506	4.490	3.51	0.412	4.690	2.91	5
6	6.625	0.908	4.700	7.09	0.736	5.065	5.93	0.602	5.349	4.97	0.491	5.584	4.13	6
7 1/8	7.125	0.976	5.056	8.20	0.792	5.446	6.86	0.648	5.751	5.75	0.528	6.006	4.78	7 1/8
8	8.625	1.182	6.119	12.01	0.958	6.594	10.05	0.784	6.963	8.42	0.639	7.270	7.00	8
10	10.750	1.473	7.627	18.66	1.194	8.219	15.61	0.977	8.679	13.09	0.796	9.062	10.87	10
12	12.750	1.747	9.046	26.25	1.417	9.746	21.97	1.159	10.293	18.41	0.944	10.749	15.29	12
13 3/8	13.375	1.832	9.491	28.88	1.486	10.225	24.18	1.216	10.797	20.26	0.991	11.274	16.84	13 3/8
14	14.000	1.918	9.934	31.64	1.556	10.701	26.50	1.273	11.301	22.20	1.037	11.802	18.44	14
16	16.000	2.192	11.353	41.33	1.778	12.231	34.60	1.455	12.915	29.00	1.185	13.488	24.09	16
18	18.000	2.466	12.772	52.31	2.000	13.760	43.79	1.636	14.532	36.69	1.333	15.174	30.48	18
20	20.000	2.740	14.191	64.58	2.222	15.289	54.05	1.818	16.146	45.30	1.481	16.860	37.63	20
22	22.000	3.014	15.610	78.14	2.444	16.819	65.40	2.000	17.760	54.82	1.630	18.544	45.56	22
24	24.000	3.288	17.029	93.00	2.667	18.346	77.85	2.182	19.374	65.24	1.778	20.231	54.21	24
26	26.000				2.889	19.875	91.36	2.364	20.988	76.57	1.926	21.917	63.62	26
28	28.000				3.111	21.405	105.95	2.545	22.605	88.78	2.074	23.603	73.78	28
30	30.000				3.333	22.934	121.62	2.727	24.219	101.92	2.222	25.289	84.69	30
32	32.000							2.909	25.833	115.97	2.370	26.976	96.35	32
34	34.000							3.091	27.447	130.93	2.519	28.660	108.81	34
36	36.000							3.273	29.061	146.80	2.667	30.346	121.98	36
42	42.000										3.111	35.405	166.00	42
48	48.000													48
54	54.000													54
63	63.000													63

Table 3-1 PE 3408 polyethylene pipe iron pipe size (IPS) pipe data, *continued*

Pipe weights are calculated in accordance with PPI TR-7 using a density of 0.955. Average inside diameter is calculated using nominal OD and minimum wall plus 6% for use in estimating fluid flows. When designing components to fit the pipe ID, refer to pipe dimensions and tolerances in applicable pipe specifications.

Pressure ratings are for water at 73.4°F. For other fluid and service temperature, ratings may differ.

Pressure Class		100 psi DR 17.0			80 psi DR 21.0			64 psi DR 26.0			51 psi DR 32.5			
IPS Pipe Size (in.)	Average D_o (in.)	Minimum Wall, t (in.)	Average D_i (in.)	Weight, w_P (lb/ft)	Minimum Wall, t (in.)	Average D_i (in.)	Weight, w_P (lb/ft)	Minimum Wall, t (in.)	Average D_i (in.)	Weight, w_P (lb/ft)	Minimum Wall, t (in.)	Average D_i (in.)	Weight, w_P (lb/ft)	IPS Pipe Size (in.)
1 1/4	1.660													1 1/4
1 1/2	1.900													1 1/2
2	2.375	0.140	2.078	0.43										2
3	3.500	0.206	3.063	0.93										3
4	4.500	0.265	3.938	1.54	0.214	4.046	1.26							4
5 3/8	5.375	0.316	4.705	2.20	0.256	4.832	1.80	0.207	4.936	1.47				5 3/8
5	5.563	0.327	4.870	2.35	0.265	5.001	1.93	0.214	5.109	1.57				5
6	6.625	0.390	5.798	3.34	0.315	5.957	2.73	0.255	6.084	2.23	0.204	6.193	1.80	6
7 1/8	7.125	0.419	6.237	3.86	0.339	6.406	3.16	0.274	6.544	2.58	0.219	6.661	2.08	7 1/8
8	8.625	0.507	7.550	5.65	0.411	7.754	4.64	0.332	7.921	3.79	0.265	8.063	3.05	8
10	10.750	0.632	9.410	8.78	0.512	9.665	7.21	0.413	9.874	5.87	0.331	10.048	4.75	10
12	12.750	0.750	11.160	12.36	0.607	11.463	10.13	0.490	11.711	8.26	0.392	11.919	6.67	12
13 3/8	13.375	0.787	11.707	13.61	0.637	12.025	11.15	0.514	12.285	9.09	0.412	12.502	7.35	13 3/8
14	14.000	0.824	12.253	14.91	0.667	12.586	12.22	0.538	12.859	9.96	0.431	13.086	8.05	14
16	16.000	0.941	14.005	19.46	0.762	14.385	15.96	0.615	14.696	13.01	0.492	14.957	10.50	16
18	18.000	1.059	15.755	24.64	0.857	16.183	20.19	0.692	16.533	16.47	0.554	16.826	13.30	18
20	20.000	1.176	17.507	30.41	0.952	17.982	24.93	0.769	18.370	20.34	0.615	18.696	16.41	20
22	22.000	1.294	19.257	36.80	1.048	19.778	30.18	0.846	20.206	24.61	0.677	20.565	19.86	22
24	24.000	1.412	21.007	43.81	1.143	21.577	35.91	0.923	22.043	29.30	0.738	22.435	23.62	24
26	26.000	1.529	22.759	51.39	1.238	23.375	42.14	1.000	23.880	34.39	0.800	24.304	27.74	26
28	28.000	1.647	24.508	59.62	1.333	25.174	48.86	1.077	25.717	39.88	0.862	26.173	32.19	28
30	30.000	1.765	26.258	68.45	1.429	26.971	56.12	1.154	27.554	45.78	0.923	28.043	36.93	30
32	32.000	1.882	28.010	77.86	1.524	28.769	63.84	1.231	29.390	52.10	0.985	29.912	42.04	32
34	34.000	2.000	29.760	87.91	1.619	30.568	72.06	1.308	31.227	58.81	1.046	31.782	47.43	34
36	36.000	2.118	31.510	98.57	1.714	32.366	80.78	1.385	33.064	65.94	1.108	33.651	53.20	36
42	42.000	2.471	36.761	134.16	2.000	37.760	109.97	1.615	38.576	89.71	1.292	39.261	72.37	42
48	48.000	2.824	42.013	175.23	2.286	43.154	143.65	1.846	44.086	117.18	1.477	44.869	94.56	48
54	54.000				2.571	48.549	181.75	2.077	49.597	148.33	1.662	50.477	119.70	54
63	63.000				3.000	56.640	247.42	2.423	57.863	201.88	1.938	58.891	162.84	63

Table 3-2 PE 3408 polyethylene pipe ductile iron pipe size (DIPS) pipe data

Pipe weights are calculated in accordance with PPI TR-7 using a density of 0.955. Average inside diameter is calculated using nominal OD and minimum wall plus 6% for use in estimating fluid flows. When designing components to fit the pipe ID, refer to pipe dimensions and tolerances in applicable pipe specifications.

Pressure ratings are for water at 73.4°F. For other fluid and service temperature, ratings may differ.

Pressure Class		200 psi DR 9.0			160 psi DR 11.0			128 psi DR 13.5			
DIPS Pipe Size (in.)	Average D_o (in.)	Minimum Wall, t (in.)	Average D_i (in.)	Weight, w_P (lb/ft)	Minimum Wall, t (in.)	Average D_i (in.)	Weight, w_P (lb/ft)	Minimum Wall, t (in.)	Average D_i (in.)	Weight, w_P (lb/ft)	DIPS Pipe Size (in.)
4	4.80	0.533	3.670	3.11	0.436	3.876	2.61	0.356	4.045	2.17	4
6	6.90	0.767	5.274	6.44	0.627	5.571	5.39	0.511	5.817	4.48	6
8	9.05	1.006	6.917	11.07	0.823	7.305	9.28	0.670	7.630	7.70	8
10	11.10	1.233	8.486	16.65	1.009	8.961	13.95	0.822	9.357	11.59	10
12	13.20	1.467	10.090	23.55	1.200	10.656	19.73	0.978	11.127	16.40	12
14	15.30	1.700	11.696	31.64	1.391	12.351	26.51	1.133	12.898	22.02	14
16	17.40	1.933	13.302	40.91	1.582	14.046	34.29	1.289	14.667	28.49	16
18	19.50	2.167	14.906	51.40	1.773	15.741	43.07	1.444	16.439	35.77	18
20	21.60	2.400	16.512	63.05	1.964	17.436	52.85	1.600	18.208	43.91	20
24	25.80	2.867	19.722	89.97	2.345	20.829	75.38	1.911	21.749	62.64	24
30	32.00				2.909	25.833	115.97	2.370	26.976	96.35	30
36	38.30							2.837	32.286	138.04	36
42	44.50										42
48	50.80										48

Pressure Class		100 psi DR 17.0			80 psi DR 21.0			64 psi DR 26.0			DIPS Pipe Size (in.)
4	4.80	0.282	4.202	1.75	0.229	4.315	1.44	0.185	4.408	1.17	4
6	6.90	0.406	6.039	3.62	0.329	6.203	2.97	0.265	6.338	2.42	6
8	9.05	0.532	7.922	6.22	0.431	8.136	5.11	0.348	8.312	4.17	8
10	11.10	0.653	9.716	9.37	0.529	9.979	7.69	0.427	10.195	6.27	10
12	13.20	0.776	11.555	13.24	0.629	11.867	10.87	0.508	12.123	8.87	12
14	15.30	0.900	13.392	17.80	0.729	13.755	14.60	0.588	14.053	11.90	14
16	17.40	1.024	15.229	23.03	0.829	15.643	18.88	0.669	15.982	15.39	16
18	19.50	1.147	17.068	28.91	0.929	17.531	23.71	0.750	17.910	19.34	18
20	21.60	1.271	18.905	35.49	1.029	19.419	29.10	0.831	19.838	23.74	20
24	25.80	1.518	22.582	50.63	1.229	23.195	41.51	0.992	23.697	33.85	24
30	32.00	1.882	28.010	77.86	1.524	28.769	63.84	1.231	29.390	52.10	30
36	38.30	2.253	33.524	111.55	1.824	34.433	91.45	1.473	35.177	74.61	36
42	44.50	2.618	38.950	150.60	2.119	40.008	123.44	1.712	40.871	100.75	42
48	50.80	2.988	44.465	196.23	2.419	45.672	160.87	1.954	46.658	131.28	48

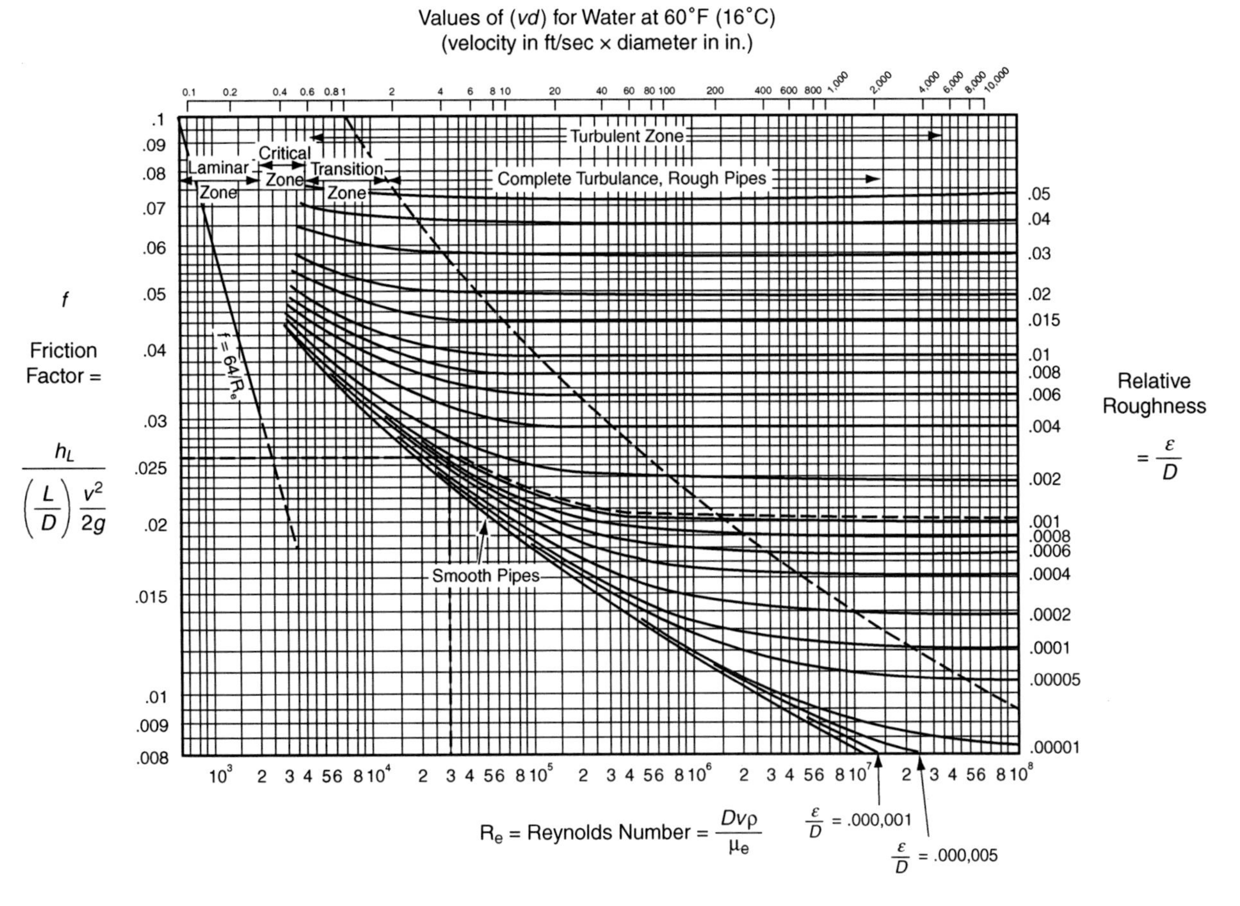

Figure 3-1 Moody diagram

The Moody diagram enables the designer to determine the flow characteristics of any fluid, gas, or liquid flowing in a particular pipe provided the surface roughness of the pipe and the viscosity and weight density of the fluid along its travel in the pipe are known. This diagram is widely accepted for hand calculations. In addition, various software programs based on the Colebrook equation and Moody diagram are available for the conducting of computerized calculations.

Example 3.1

Find the head loss and pressure drop in 1,000 ft of 10 in. DR17 PE pipe (IPS) carrying water at a velocity of 5 ft/sec.

First, estimate the Reynolds number from the Moody diagram. Multiply the velocity times the pipe average, D_i, to get the (vD_i) term, which equals 9.410 in. × 5 ft/sec or 47.1 in.-ft/sec. Look across the top of the chart and then directly down to the lower abscissa to find the Reynolds number for water. Re = 3.5 × 10^5. The inside diameter is obtained from Tables 3-1 and 3-2.

The relative pipe roughness for 10 in. PE pipe is:

$$\frac{\varepsilon}{D} = \frac{0.000005 \text{ ft}}{9.410 \text{ in.}} = 0.00000064$$

This value of relative roughness corresponds with the "smooth pipes" line on the diagram. Draw a vertical line from the Reynolds number of 3.5 × 10^5 upward until it intersects the "smooth pipes" line on the diagram. The friction factor, f, is given on the left side of the diagram as 0.014. Eq 3-3 can be solved for the head loss.

$$h_f = f\frac{v^2 L}{2gD} = 0.014\frac{5^2}{2(32.2)}\frac{1{,}000}{(9.410/12)} = 6.93 \text{ ft of head}$$

The corresponding pressure drop, p_f, equals 6.93/2.31 or 3.0 psi. Moody diagram calculations are assumed accurate to ±10 percent.

HAZEN-WILLIAMS FORMULA

Over the years, various empirical formulas have been proposed for the more direct and simpler computing of frictional losses incurred in pipes that convey water. First proposed in 1904 and still very widely used is the Hazen-Williams formula, which is based on water at 60°F:

$$h_f = 0.002083L\left(\frac{100}{C}\right)^{1.85}\left(\frac{Q^{1.85}}{D_i^{4.87}}\right) \quad \text{(Eq 3-8)}$$

Where:

h_f = friction loss, feet of water
L = length of pipe, ft
C = Hazen-Williams pipe flow coefficient, dimensionless
Q = volumetric flow rate, gpm
D_i = pipe inside diameter, in.

The flow velocity is given by:

$$v = 0.115CD_i^{0.63}s^{0.54} \quad \text{(Eq 3-9)}$$

Where:

v = flow velocity, ft/sec
C = Hazen-Williams pipe flow coefficient, dimensionless
D_i = pipe inside diameter, in.
s = hydraulic slope (the resultant friction loss in feet of water per foot of pipe), ft/ft

By combining Eq 3-9 with Eq 3-4 and when expressing the inside pipe diameter in inches rather than feet, the following commonly used formula is obtained for the flow rate in a pipeline:

Table 3-3 Representative equivalent length in pipe diameters of various piping components

Piping Component	Equivalent Length in Pipe Diameters (L_{eq}/D_i)
90° Molded elbow	40
45° Molded elbow	21
90° Fabricated elbow	32
75° Fabricated elbow	27
60° Fabricated elbow	21
45° Fabricated elbow	16
30° Fabricated elbow	11
15° Fabricated elbow	5
Equal outlet tee, run/branch	80
Equal outlet tee, run/run	27
Globe valve, conventional, fully open	340
Angle valve, conventional, fully open	145
Butterfly valve, ≥ 8-in., fully open	40
Check valve, conventional swing	135
Insert couplings	12
Male-female insert adapters	18

$$Q = 0.2815 C D_i^{2.63} s^{0.54} \quad \text{(Eq 3-10)}$$

The appropriate value of C to be used in the Hazen-Williams equation depends to a large extent on the roughness of the pipe. But, to a lesser extent, it is also affected by pipe diameter, water velocity, and water temperature. Hydraulic studies show that the flow disturbance effect of PE butt-fused joints is minimal and that the Hazen-Williams flow coefficient for butt-fused PE pipes is around 155. The common practice is to assume a C coefficient of 150 for PE pipelines. No allowance for corrosion and therefore, no subsequent lowering of the flow capacity need be considered when using PE pipe.

FITTINGS

Friction head losses that develop in piping components such as elbows, tees, and vessel connections are generally expressed as loss of static head in feet of fluid flowing, or as an equivalent length (L_{eq}) of straight pipe of the same size that would produce the same static head loss at the flow conditions for which the pipeline is designed. For the various sizes of a line of products such as fittings and valves that are of a consistent design, the test data indicate that the ratio of equivalent pipe length (L_{eq}) to inside pipe diameter (D_i) tends to be a constant value. Representative values of equivalent pipe length in terms of the number of pipe diameters are listed in Table 3-3 for components frequently used in PE piping systems.

When losses are expressed as equivalent pipe diameters, it is a simple matter to multiply the value by the nominal pipe diameter to obtain the equivalent pipe length represented by the head loss induced by the component. After the equivalents of all components in a pipeline are determined, they must be added to straight pipe length before computing the total head loss.

AIR BINDING

In rolling or mountainous country, additional drag caused by air binding must be avoided. Air binding occurs when air in the system accumulates at local high spots. This reduces the effective pipe bore and restricts flow. Vents, such as standpipes or air release valves, may be installed at high points to avoid air binding. If the pipeline has a high point above that of either end, vacuum venting may be required to prevent vacuum collapse, siphoning, or to allow drainage.

Rapid venting of a pipeline may have potential adverse consequences. In some cases, entrapped air that is rapidly vented can greatly raise fluid pressures causing damage to pipe and/or pipe components.

Example 3.2

Determine the size of pipe required to handle a flow of 680 gpm at a pressure of 100 psi (80°F [27°C]). Find the pressure drop over 1,000 ft of this pipeline. What is the resulting velocity?

Because the line is operating at 100 psi and 80°F (27°C), DR17 should be adequate for handling the pressure. As the Hazen-Williams equation cannot be solved directly for diameter if only the flow rate is known, the normal procedure is to assume a diameter and check if the pressure drop and velocity are acceptable.

Start by assuming a nominal DIPS diameter of 8 in. The ID of 8-in DR17 DIPS pipe is given in Table 3-2 and equals 7.922 in. The resulting friction loss is given by Eq 3-8:

$$h_f = (0.002083)(1{,}000)\left(\frac{100}{150}\right)^{1.85}\left(\frac{680^{1.85}}{7.922^{4.87}}\right) = 7.17 \text{ ft of head}$$

The pressure drop per 1,000 ft in psi equals:

$$7.17 \text{ ft of head} \times \frac{1 \text{ psi}}{2.31 \text{ ft of head}} = 3.2 \text{ psi}$$

Flow velocity is given by Eq 3-9. Because s is the feet of water head per foot of pipe, s equals 0.00717.

$$v = 0.115(150)(7.922)^{0.63}(0.00717)^{0.54} = 4.42 \text{ ft/sec}$$

Thus, 8-in DR17 DIPS pipe can handle a flow of 680 gpm at a 100 psi pumping pressure.

Chapter 4

Working Pressure Rating

INTRODUCTION

The parameters that control the design of PE pipe for water distribution and transmission are the sustained internal pressure, the intermittent or transient surges that may occur during operation, and the temperature of operation. The working pressure rating (*WPR*) accounts for all of these factors. This chapter covers the current design guidelines for determining the working pressure rating of a PE water distribution and transmission system made to ANSI/AWWA C901[1] and C906[2].

The definitions for pressure class, recurring and occasional surge, working pressure and working pressure rated for polyethylene pipe are provided as an introduction to this section for reference.

Pressure class *(PC)*. The pressure class is the design capacity to resist working pressure up to 80°F (27°C) maximum service temperature with specified maximum allowances for recurring positive pressure surges above working pressure.

Surge pressure *(PS)*. Surge pressure is the maximum hydraulic transient pressure increase (sometimes called water hammer) in excess of the operating pressure that is anticipated in the system as the result of sudden changes in velocity of the water column.

For purposes of product selection and design, the following surge definitions are applied:

Recurring surge pressure *(P_{RS})*. Recurring surge pressures occur frequently and are inherent to the design and operation of the system (such as, normal pump start-up or shut-down and normal valve opening or closure).

Occasional surge pressure *(P_{OS})*. Occasional surge pressures are caused by emergency operations. Occasional surge pressures are usually the result of a malfunction, such as a power failure or system component failure, which includes pump seize-up, valve-stem failure, and pressure-relief valve-failure.

Working pressure *(WP)*. Working pressure is the maximum anticipated, sustained operating pressure applied to the pipe exclusive of transient pressures.

Working pressure rating *(WPR)*. The working pressure rating is the design capacity to resist working pressure at the anticipated operating temperature with sufficient

Table 4-1 Hydrostatic design basis (*HDB*) for standard PE 3408 and PE 2406 materials*

Service Temperature	PE 3408 *HDB*	PE 2406 *HDB*
73°F (23°C)	1,600 psi (11.03 MPa)	1,250 psi (8.62 MPa)

* Consult PPI TR-4 for elevated temperature *HDB* ratings.

capacity against the actual anticipated positive pressure surges above working pressure. A pipe's *WPR* may be equal to, or less than, its nominal *PC*.

For most design situations, the *PC* offers abundant margin for surge. In those few cases where the *PSs* may be excessive, the *WPR* may be reduced to accommodate the additional surge. The design guidelines for determining the *WPR* are provided in this chapter.

PRESSURE CLASS

The sustained internal pressure, exclusive of transient pressure surges, that can be applied to a pipe system is its *PC,* which is a function of the allowable hoop stress and pipe thickness. The allowable hoop stress is determined by testing pipe in accordance with ASTM D2837[3] to obtain a hydrostatic design basis (*HDB*). The technique for developing an *HDB* rating is outlined in Chapter 1. PE materials are typically rated at two temperatures, 73°F (23°C) and 140°F (60°C). *HDB* ratings for typical PE materials used for water service are provided in Table 4-1. A periodically updated listing of *HDBs* for commercial grades of PE is published in PPI TR-4[4].

The internal pressure capability or *PC* of a PE pipe can be calculated from the *HDB* as a function of wall thickness and pipe diameter using the ISO equation and the appropriate design factor. The ISO equation is provided in Equations 4-1a–4-1d in terms of *DR,* outside diameter and wall thickness, *IDR,* and inside diameter and wall thickness.

ISO equation in terms of *DR*

$$PC = \frac{2(HDB)(DF)}{DR - 1} \qquad \text{(Eq 4-1a)}$$

ISO equation in terms of *OD* and wall thickness

$$PC = \frac{2t(HDB)(DF)}{D_o - t} \qquad \text{(Eq 4-1b)}$$

ISO equation in terms of *IDR*

$$PC = \frac{2(HDB)(DF)}{IDR + 1} \qquad \text{(Eq 4-1c)}$$

ISO equation in terms of *ID* and wall thickness

$$PC = \frac{2t(HDB)(DF)}{D_i + t} \qquad \text{(Eq 4-1d)}$$

Where:

PC = pressure class, psig

HDB = hydrostatic design basis, psi

DF = design factor (dimensionless), 0.5 for water applications

DR = dimension ratio (dimensionless)

$$= \frac{D_o}{t}$$

D_o = average outside diameter, in.

t = minimum wall thickness, in.

IDR = inside (diameter) dimension ratio (dimensionless)

$$= \frac{D_i}{t}$$

D_i = average inside diameter, in.

For PE pipe applications where water pressure is applied continuously, a DF of 0.5 is generally accepted. This factor takes into consideration the complex loadings a pipe may experience during service as a result of installation quality and service conditions. The result is a maximum hydrostatic pressure that can be continuously applied with a high degree of certainty that failure of the pipe will not occur over the service life.

Using the previous equations and the DF of 0.5 for pipes produced with standard grade PE 3408 or PE 2406 materials and operated at service temperatures through 80°F (27°C), the calculations result in the pressure classes given in Table 4-2.

SURGE CONSIDERATIONS

When there is a sudden increase or decrease in flow velocity in a pipe system, a pressure surge will occur. This type of event is called a *hydraulic transient* or *water hammer*. The transient pressure is a rapidly moving wave that increases and decreases the pressure in the system depending on the source of the transient and the direction of the wave. Water hammer can be caused by rapid opening or closing of valves and hydrants, start-up or shut-down of pumps, sudden loss of power, or any other situation that causes a sudden change in velocity of the fluid stream. These surges are either recurring or occasional as defined in ANSI/AWWA C906[3]:

Recurring Surge Pressure (P_{RS})

Recurring surge pressures occur frequently and are inherent in the design and operation of the system. Recurring surge pressures may be caused by normal pump start-up or shut-down and normal valve opening or closure. P_{RS} is the allowance for recurring surge pressure (Eq 4-4).

Occasional Surge Pressure (P_{OS})

Occasional surge pressures are caused by emergency operations. Occasional surge pressures are often the result of a malfunction, such as a power failure or system component failure, including pump seize-up, valve-stem failure, and pressure-relief valve-failure. P_{OS} is the allowance for occasional surge pressure (Eq 4-5).

PE pipes can safely withstand momentarily applied maximum pressures that are significantly above the pipe's PC because of the viscoelastic nature of the material.

Table 4-2 Pressure class* for PE 3408 and PE 2406 pipe

	PE 3408 PC		PE 2406 PC	
Pipe DR/IDR	*psig*	*MPa*	*psig*	*MPa*
7.3/5.3	254	1.75	198	1.37
9.0/7.0	200	1.38	156	1.08
9.3/7.3	193	1.33	151	1.04
11.0/9.0	160	1.10	125	0.86
13.5/11.5	128	0.88	100	0.69
15.5/13.5	110	0.76	86	0.59
17.0/15.0	100	0.69	78	0.54
21.0/19.0	80	0.55	63	0.43
26.0/24.0	64	0.44	50	0.34
32.5/30.5	51	0.35	40	0.28

* Pressure class is applicable to operating temperatures through 80°F (27°C). For temperatures above 80°F (27°C), a temperature compensation multiplier, F_T, must be applied. See Working Pressure Rating section. Allowances for surge events are presented in the Surge Considerations section.

The strain from an occasional, limited load of short duration is met with an elastic response that is reversed on the removal of the load. This temporary elastic strain causes no damage to the pipe material and has no adverse effect on the pipe's long-term strength as confirmed by experiment[5].

In most dynamic pumping systems, surge pressures may occur repeatedly. A satisfactory piping material must have sufficient fatigue resistance to withstand repeated surges. PE piping materials are highly fatigue resistant. The high molecular weight materials specified in ANSI/AWWA C901[1] and C906[2] provide very high fatigue endurance. Tests conducted on PE pipe document a capacity to resist recurring transient pressure surges that can rise significantly above the sustained pressure rating of the piping[6,7,8]. Negative pressure should be checked in accordance with the section Wall Buckling in Chapter 5. Negative pressure greater than 1.0 atmosphere is not recommended as column separation in the pipeline will occur and the resulting positive pressure cannot be predicted.

Pressure Surge Calculations

The following method may be used to estimate the transient pressure surge, or water hammer, that may occur in a PE pipeline conveying water.

An abrupt change in the velocity of a flowing liquid generates a pressure wave. The velocity of which is given by the following equation:

$$a = \frac{4{,}660}{\sqrt{1 + \frac{K}{E_d}(DR - 2)}} \quad \text{(Eq 4-2)}$$

Where:

a = wave velocity (celerity), fps

K = bulk modulus of fluid at working temperature (300,000 psi for water at 73°F)

E_d = dynamic instantaneous effective modulus of pipe material (150,000 psi for PE pipe)

DR = pipe dimension ratio

The resultant transient surge pressure may be calculated from the wave velocity and the change in fluid velocity:

$$P_s = a\left(\frac{\Delta v}{2.31g}\right) \quad \text{(Eq 4-3)}$$

Where:

P_s = transient surge pressure, psig

a = wave velocity (celerity), fps

Δv = velocity change occurring within the critical time $2L/a$, in seconds, where L is the pipe length, ft

g = gravitational acceleration, 32.2 ft/sec^2

Table 4-3 provides calculated values for the surge pressure in PE pipes caused by sudden velocity changes of 1 ft/sec and 5 ft/sec. To determine the surge pressure in a PE pipeline for a different sudden velocity change, multiply the Table 4-3 surge pressure for 1.0 velocity change for the corresponding *DR* by the desired value of the sudden velocity change. For example, the surge pressure in a *DR* 17 pipe subjected to a sudden velocity change of 3.5 fps is 3.5 × 11.3 = 39.6 psig.

The magnitude of the surge pressure depends on the stiffness of the pipe material as well as the pipe dimensions. Greater stiffness yields greater wave velocity and higher surge pressure.

The *PC* defined in ANSI/AWWA C906 includes surge pressure allowances for recurrent and occasional water hammer events. Surge pressure allowances are applied above the *PC*. For recurrent surges, the allowance is 50 percent of the *PC*, and for occasional surges, the allowance is 100 percent of the *PC*. A PE piping system that is subjected to recurring surges over its service life will experience a greater number of surge events compared to a system that is subjected to occasional surges. For this reason, the allowance for recurring surges is lower than that for occasional surges.

$$P_{RS} = 0.5 \times PC \quad \text{(Eq 4-4)}$$

$$P_{OS} = 1.0 \times PC \quad \text{(Eq 4-5)}$$

Where:

P_{RS} = pressure allowance for recurrent surge, psig

P_{OS} = pressure allowance for occasional surge, psig

The pressure allowance for surge is applied exclusively to pressure that occurs during a surge event, never to sustained operating pressure. When a surge event occurs in a PE piping system, the maximum allowable pressures, sustained operating pressure plus surge allowance pressure are described in Eq 4-6 and 4-7.

Table 4-3 Surge pressures generated by a sudden change in water flow velocity for PE pipes operating at service temperatures through 80°F (27°C)

Pipe DR	Surge Pressure for Sudden Velocity Change	
	1.0 fps Δv Velocity Change (psig)	5.0 fps Δv Velocity Change (psig)
32.5	8.0	39.8
26	9.0	44.8
21	10.0	50.2
17	11.3	56.3
13.5	12.8	63.9
11	14.4	71.9
9	16.2	80.9
7.3	18.4	91.9

For a recurrent surge event:

$$P_{(MAX)(RS)} = PC + P_{RS} \quad \text{(Eq 4-6)}$$

For an occasional surge event:

$$P_{(MAX)(OS)} = PC + P_{OS} \quad \text{(Eq 4-7)}$$

Where:

$P_{(MAX)(RS)}$ = maximum allowable system pressure during recurrent surge, psig

$P_{(MAX)(OS)}$ = maximum allowable system pressure during occasional surge, psig

Tables 4-4 and 4-5 show *PC* surge pressure allowances and corresponding sudden changes in flow velocity for PE 3408 and PE 2406 pipe, respectively.

WORKING PRESSURE RATING

AWWA standards for PE pipe define *WPR* as the capacity to resist *WP* with sufficient capacity against the actual anticipated positive pressure surges above working pressure. *WP* is defined as the maximum anticipated, sustained operating pressure applied to the pipe exclusive of transient pressures.

The *PC* defines the limits of the pipe for sustained and surge pressures at operational temperatures through 80°F (27°C). Where the operating temperature is above 80°F (27°C) or surge pressures are expected to be higher than those allowed by the *PC* definition, or if both conditions apply, the *WPR* must be reduced below the *PC*. The *WPR* can never exceed the *PC*. *WP*, *WPR*, and *PC* are related as follows:

$$WP \leq WPR \leq PC \quad \text{(Eq 4-8)}$$

Table 4-4 Pressure class, surge allowance, and corresponding sudden velocity change for PE 3408 pipe operating at service temperatures through 80°F (27°C)

		Recurring Surge Events		Occasional Surge Events	
DR	PC (*psi*)	P_{RS} (*psi*)	Corresponding Sudden Velocity Change (*fps*)	P_{OS} (*psi*)	Corresponding Sudden Velocity Change (*fps*)
7.3	254	127	6.9	254	13.8
9	200	100	6.2	200	12.4
9.3	193	96	6.1	193	12.2
11	160	80	5.6	160	11.1
13.5	128	64	5.0	128	10.0
15.5	110	55	4.7	110	9.3
17	100	50	4.4	100	8.9
21	80	40	4.0	80	8.0
26	64	32	3.6	64	7.2
32.5	51	25	3.2	51	6.4

Table 4-5 Pressure class, surge allowance, and corresponding sudden velocity change for PE 2406 pipe operating at service temperatures through 80°F (27°C)

		Recurring Surge Events		Occasional Surge Events	
DR	PC (*psi*)	P_{RS} (*psi*)	Corresponding Sudden Velocity Change (*fps*)	P_{OS} (*psi*)	Corresponding Sudden Velocity Change (*fps*)
7.3	198	99	5.4	198	10.8
9	156	78	4.8	156	9.7
9.3	151	75	4.7	151	9.5
11	125	63	4.3	125	8.7
13.5	100	50	3.9	100	7.8
15.5	86	43	3.6	86	7.3
17	78	39	3.5	78	6.9
21	63	31	3.1	63	6.2
26	50	25	2.8	50	5.6
32.5	40	20	2.5	40	5.0

When PE pipe operates at 80°F (27°C) or less and the expected recurrent or occasional surge pressures are within the limits established by Eq 4-4 and 4-5, respectively, the *WPR* equals the *PC*:

$$WPR = PC \qquad \text{(Eq 4-9)}$$

When PE pipe operates at temperatures above 80°F (27°C), a temperature compensation multiplier, F_T, is applied to determine the *WPR*:

$$WPR = PC \times F_T \quad \text{(Eq 4-10)}$$

Temperature compensation multipliers, F_T, for PE pipe maximum operating temperatures above 80°F through 100°F are presented in Table 4-6.

WPR must also be evaluated to account for the expected recurrent or occasional surges (P_{OS}) within the pipe system.

For recurring surges, *WPR* is one and one half times the pipe's *PC* adjusted for temperature, less the maximum pressure allowance resulting from recurring pressure surges (P_{RS}):

$$WPR = 1.5(PC)(F_T) - P_{RS} \quad \text{(Eq 4-11)}$$

For occasional surge, *WPR* is two times the pipe's *PC* adjusted for temperature, less the maximum pressure allowance resulting from occasional pressure surges (P_{OS}):

$$WPR = 2.0(PC)(F_T) - P_{OS} \quad \text{(Eq 4-12)}$$

The *WPR* is the smallest number determined in accordance with Eq 4-10, 4-11, and 4-12.

Operating at a *WP* that is less than the pipe's *PC* provides additional capacity for surge pressure. However, surge allowance is applied exclusively for surge events and is never used to increase *WP*.

Using Eq 4-11, *WPR* values for a PE 3408 PE pipe meeting the requirements of ANSI/AWWA C906 as a function of the velocity change within the pipeline were determined and are presented in Table 4-7. The shaded areas indicate conditions where the *WPR* is equivalent to the *WP*; *WPR* = *PC*.

Pipe Design Examples

1. Calculate the *WPR* for an 8-in. IPS *DR* 17 pipe transporting water at a maximum velocity of 4.4 fps and operating at 73°F (23°C).

   Using Table 4-7, the *WPR* would be equal to the *PC* of 100 psig.

2. Calculate the *WPR* for an 8-in IPS *DR* 17 pipe transporting water and operating at 73°F (23°C) where the system will have recurring surges and a maximum flow velocity of 5.0 fps.

   Using Eq 4-11 and the appropriate P_{RS} value from Table 4-3, the calculation is as follows:

$$WPR = (1.5)(100)(1.0) - 56.3 = 93.7 \text{ psi}$$

   This is in agreement with the value in Table 4-7.

3. Calculate the *WPR* for an 8-in. IPS *DR* 17 pipe operating at 100°F (38°C) where the water velocity will occasionally reach 10 fps.

Table 4-6 Temperature compensation multipliers, F_T

Maximum Operating Temperature		Temperature Compensation Multiplier, F_T
°F	°C	
Below 81°F	Below 28°C	1.0
From 81°F to 90°F	From 28°C to 32°C	0.9
From 91°F to 100°F	From 33°C to 38°C	0.8
Above 100°F	Above 38°C	—

NOTE: The upper operating temperature limit, as well as the temperature compensation multiplier for temperatures above 100°F (38°C), can vary depending on the pipe material. The pipe manufacturer should be consulted for this information.

Table 4-7 PE 3408 working pressure rating for recurring surge events as a result of instantaneous change in water column velocity

	WPR (*psi*) vs. Recurring Instantaneous Change in Water Column Velocity (*fps*)						
Pipe DR	0-2 *fps*	3 *fps*	4 *fps*	5 *fps*	6 *fps*	7 *fps*	8 *fps*
7.3	254	254	254	254	254	252	234
9	200	200	200	200	200	187	171
9.3	193	193	193	193	193	178	162
11	160	160	160	160	154	139	125
13.5	128	128	128	128	115	102	90
15.5	110	110	110	106	94	83	71
17	100	100	100	94	82	71	60
21	80	80	80	70	60	50	40
26	64	64	60	51	42	33	24
32.5	51	51	44	36	28	20	13

NOTE: While the pipe is adequate to resist the positive pressure shown in the shaded areas of the table, negative pressure must be prevented from exceeding 1.0 atmosphere to prevent water column separation.

Using Eq 4-12, the 1.0 fps velocity change surge pressure from Table 4-3, and the 100°F temperature compensation multiplier, F_T, from Table 4-6, the calculation is as follows:

$$WPR = (2)(100)(0.8) - 10(11.3) = 43 \text{ psi}$$

MOLDED AND FABRICATED FITTINGS

The *WP* of a molded or fabricated fitting is established using long-term sustained pressure tests in a manner similar to that used for pipe. The long-term strength of each member of a family of fittings made from a particular PE material is established from the results of long-term pressure tests conducted on products that represent each kind of fitting and the size range in which that fitting is available. The *PC* of each fitting is calculated from the results of these tests.

Molded fittings are commercially available with a *PC* up to 254 psig (in some cases higher) for water at 73°F (23°C). Fabricated fittings are custom-made to pressure classes from 40 to 200 psig or higher depending on the size and manufacturer. In either case, fittings shall meet the requirements of ANSI/AWWA C901 or C906.

REFERENCES

1. ANSI/AWWA C901, *Polyethylene (PE) Pressure Pipe and Tubing ½ In.(13 mm) Through 3 In. (76 mm) for Water Service,* American Water Works Association, Denver, CO.
2. ANSI/AWWA C906, *Polyethylene (PE) Pressure Pipe and Fittings, 4 In.(100 mm) Through 63 In. (1,575 mm), for Water Distribution and Transmission,* American Water Works Association, Denver, CO.
3. ASTM D2837, *Standard Test Method for Obtaining Hydrostatic Design Basis for Thermoplastic Pipe Materials or Pressure Design Basis for Thermoplastic Pipe Products,* ASTM International, West Conshohocken, PA.
4. PPI TR-4, *PPI Listing of Hydrostatic Design Basis (HDB), Strength Design Basis (SDB), Pressure Design Basis (PDB) and Minimum Required Strength (MRS) Ratings for Thermoplastic Piping Materials or Pipe,* Plastics Pipe Institute, Washington, DC.
5. Szpak, E. and Rice, F.G., 1977. "Strength of Polyethylene Piping – New Insights," Parts I and II, *Engineering Digest.*
6. Fedossof, F.A. and E. Szpak, 1978. "Cyclic Pressure Effects on High Density Polyethylene Pipe," a paper presented at the Western Canada Sewage Conference, Regina, Sask., Canada.
7. Bowman, J.A., "The Fatigue Response of Polyvinyl Chloride and Polyethylene Pipe Systems," *Buried Plastics Pipe Technology,* ASTM STP1093, American Society for Testing and Materials, Philadelphia, 1990.
8. Janson, L.E., 1996. *Plastic Pipes for Water Supply and Sewage Disposal.* Borealis, Sven Axelson AB/Affisch & Reklamtryck AB, Boras, Sweden.

Chapter 5

External Load Design

INTRODUCTION

A buried pipe that is used to convey fluids under pressure must safely withstand all those stresses and strains that can result from two kinds of loadings: the *internal* loads caused by working fluid pressure and fluid flow dynamics (e.g., water hammer) and the *external* loads that usually include the backfill weight (dead load) and the earth pressure from vehicles (live load) but can also include additional pressures resulting from groundwater, vacuum pressure (e.g., external atmospheric pressure), and surface structures. While internal loads develop tensile stresses in the axial (longitudinal) and ring (circumferential) directions, external loadings principally develop ring bending and ring compressive stresses. The ring bending stresses are a reaction to the external loads, which deform the pipe from its initial round shape.

Pipes intended for buried applications are commonly differentiated into two classes, *rigid* and *flexible,* depending on their capacity to deform in service without cracking, or otherwise failing. PE pipe can safely withstand considerable deformation and is therefore classified as a flexible pipe.

Flexible pipe's capacity to deform significantly without undergoing structural damage allows the development of a soil/pipe interaction that, because of the mechanics of the soil/pipe structure, works to shed some of the backfill load caused by arching, further compresses the soil at the side of the pipe that enhances lateral support, and reduces (and ultimately limits) further pipe deformation. While all flexible pipes should have sufficient compressive strength to safely withstand the compressive stresses generated by the external loadings, their usual design constraint is not material failure caused by compression but rather the ensuring that the earth and groundwater pressures will not excessively deflect or buckle the pipe.

Soil arching reduces the external load on a flexible pipe to less than the stress caused by the burden of soil directly above the pipe ("geostatic" stress). This reduction results in flexible pipes having lower bending and compressive stresses than rigid pipes. And because PE is a viscoelastic material, the ring bending stress decreases over time[1]. When plastic is essentially under constant strain, the load required to maintain the constant deformation gradually decreases, and after a sufficient length of time, it can be as low as a fourth of the initial value.

When a buried flexible pipe is pressurized internally, the compressive and bending stresses caused by external loads are reduced as the pipe rerounds. Furthermore, the design factors (safety factors) commonly used for design of pipe against internal fluid pressure give some consideration to the presence of other stresses than those resulting from the fluid pressure.

For the reasons stated previously, a combined internal and external loading analysis is not necessary for PE pipe. The accepted convention is to design PE pipe as if internal and external loads act independently. Most often, pressure design is the controlling factor. Generally, the design procedure is to select a pipe that satisfies the internal working pressure, maximum anticipated surge, and flow capacity requirements and then to analyze the subsurface installation to ensure that the pipe, as installed, will withstand the external loads. External load capacity is normally checked by the following steps:

1. Determine the external loads applied to pipe.
2. Verify that the deflection caused by external loads does not exceed the allowable value.
3. Verify that the combined pressure caused by external loads and internal vacuum is less than the allowable buckling pressure of the pipe.
4. Verify that the compressive stress in the pipe wall caused by external loads does not exceed the allowable value.

For convenience, the design window section includes DR7.3 through DR21 HDPE pipes. Pipes installed according to the design window meet the specified deflection limits, have a safety factor of at least 2 against buckling, and do not exceed the allowable compressive stress. Thus, the designer need not perform calculations on pipes within the design window.

NOTE: This chapter is limited to loading on pipes buried in trenches. The load and pipe reaction calculations may not apply to pipes installed by trenchless technologies, such as pipe bursting and directional drilling. Directional drilled pipe typically does not have side support as provided by the embedment for pipe installed in trenches. See ASTM F1962.

DEAD LOADS

The load applied to a buried pipe caused by the weight of soil above the pipe depends on the relative flexibility between the pipe and the surrounding embedment and in-situ soil. Where the pipe is flexible, its crown will tend to deflect downward and permit arching to occur in the backfill. Arching tends to transfer some of the weight of the backfill to the soil beside the pipe and thus reduces the load on the pipe. Marston's equation gives the earth load considering arching[2].

A practical and conservative approach for designing PE pipe is to ignore arching and assume that the dead load on the pipe equals the weight of the column (or prism) of soil directly above the pipe. This load is referred to as the *prism* load and is given in Eq 5-1 as vertical earth pressure acting on the horizontal surface at the pipe crown.

$$P_E = wH \qquad \text{(Eq 5-1)}$$

Where:

P_E = earth pressure on pipe, psf

w = unit weight of soil, pcf

H = soil height above pipe crown, ft

For depths of cover in excess of 50 ft or more, precise calculation of earth load on the pipe should be conducted because the benefits of arching on load reduction can be quite significant. See Burns and Richards[3].

LIVE LOADS

Wheel loads from trucks or other vehicles are significant for pipe buried at shallow depths. The resultant load transferred to the pipe depends on the vehicle weight, the tire pressure and size, vehicle speed, surface smoothness, pavement, and distance from the pipe to the point of loading.

To develop reasonably well-distributed earth pressures, pipe subjected to vehicular loading should be installed at least 18 in. or one pipe diameter (whichever is larger) under the road surface. When this condition cannot be met, an analysis addressing the stability of the soil envelope and bending stiffness of the pipe crown is required. See Moore[4] or Watkins/Anderson[5].

AASHTO H20 Wheel Loads

The most common loading used for design is H20 highway loading. The American Association of State Highway and Transportation Officials (AASHTO) publishes wheel loadings for standard trucks. A standard H20 truck has a front axle load of 8,000 pounds and a rear axle load of 32,000 pounds, for a total weight of 40,000 pounds or 20 tons. A tractor truck with semitrailer has the same wheel loading as an H20 truck. The loading for a semitrailer and tractor is designated HS20 and is shown in Figure 5-1. At the rear axle(s), each wheel load is 0.4 W, where W is the total weight of the truck. The 0.4 W wheel load may be used to represent the load applied by either a single axle or tandem axle. The imprint (contact) area for dual tires on an H20 truck is typically taken as 10 in. by 20 in. The heaviest tandem axle loads normally encountered on highways are around 40,000 pounds. Occasionally, vehicles may be permitted with loads up to 50 percent higher.

The standard AASHTO wheel loading is a static load. However, when a vehicle in motion strikes bumps or other imperfections in the road surface the resulting impact increases the downward force. The resulting load is generally found by multiplying the static load by an "impact factor." The impact factor depends on the depth of cover. Some states permit a reduction in this factor with depth. See Table 5-1 for typical impact factors for paved roads. Off-highway vehicles may be considerably heavier than H20 trucks, and these vehicles frequently operate on unpaved roads, which may have uneven surfaces. On unpaved roads, impact factors of 2.0, 3.0, or higher may occur depending on vehicle speed.

Highway Load (Rigid Pavement)

For AASHTO H20 highway vehicular loading through rigid pavement, the soil pressure acting at the crown of the pipe can be obtained from Table 5-2. These values were developed by the American Iron and Steel Institute (AISI) and are given in ASTM A796[6]. The H20 highway loading assumes a 16,000 lb wheel load applied through a 12-in. thick pavement. For use with heavier trucks, the table can be adjusted proportionally to the increased weight as long as the truck has the same tire area (10 in. by 20 in.) as an H20 truck. So, the values for a 64,000 lb axle load vehicle having the same tire contact area as an H20 truck can be obtained by doubling the value in the AISI table.

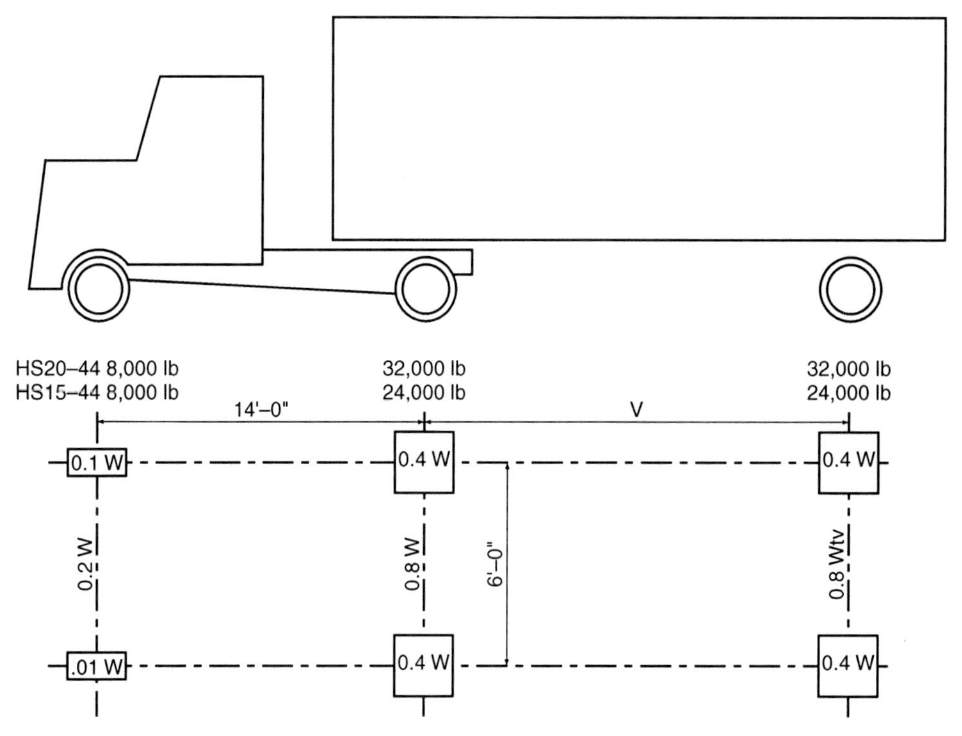

Figure 5-1 AASHTO HS20 wheel load distribution

Table 5-1 Impact factors for paved road

Impact Factor, I_f	Cover Depth, *H (ft)*
1.50	2.0 < H < 3.0
1.40	3.0 < H < 4.0
1.30	4.0 < H < 5.0
1.20	5.0 < H < 6.0
1.10	6.0 < H <7.0
1.00	H > 7.0

Flexible Pavement and Unpaved Roads

Flexible pavement transmits considerably more concentrated live load pressure than rigid pavement. The loadings expressed in Table 5-2 are not appropriate for H20 trucks on flexible pavement or on unpaved surfaces. Stewart and O'Rourke[7] show that rigid pavement significantly reduced live load compared to flexible or very flexible pavements. The live load acting on a pipe buried under flexible pavement or an unpaved road can be found by integrating Boussinesq's point load equation over the pipe's diameter. See Holtz and Kovacs[8]. Table 5-3 gives live load pressures based on Boussinesq's integrations for 4-in. diameter pipe. These values are conservative for use with larger diameter pipe, and any error incurred with their use for smaller diameter pipe is insignificant. The integrations were done for both a single vehicle load and two passing vehicles at a distance of 3.0 ft between wheels and considering pipe both parallel

Table 5-2 H20 loading (rigid pavement[6])

Height of Cover *(ft)*	Load *(psi)*
1.5*	9.7
2	5.6
3	4.2
4	2.8
5	1.7
6	1.4
7	1.2
8	0.7

* The AISI table includes load at 1.0 ft of cover and does not give the load at 1.5 ft. The value at 1.5 ft was added to this table using the same method of calculation as used in the AISI table with an impact factor of 1.2.

Table 5-3 AASHTO H20 loading under flexible pavement and unpaved roads

Height of Cover *(ft)*	Load *(psi)*
1.5	13.9
2.0	9.5
2.5	7.0
3.0	5.4
3.5	4.3
4	3.6
6	2.0
8	1.3
10	0.8

and transverse to the direction of traffic. The largest values from these integrations were used in the table.

Off-Road Vehicle Loads

Some vehicles, such as off-road haul trucks, have a footprint that does not match an H20 vehicle. Timeoshenko's equation[9] for calculating soil pressure directly beneath a concentrated load acting over a specified area may be used to find the liveload.

$$P_L = \frac{I_f W_L}{144\,A}\left(1 - \frac{H^3}{(R^2 + H^2)^{1.5}}\right) \qquad \text{(Eq 5-2)}$$

Where:

P_L = vertical stress acting on the pipe crown, psi

I_f = impact factor

W_L = wheel load, lbs

A = contact area, ft^2

R = equivalent radius, ft

H = soil height above pipe crown, ft

For standard H20 highway vehicle loading, the contact area is normally taken for dual wheels as 16,000 lb over a 10 in. by 20 in. area, which equals 1.4 ft^2. The equivalent radius is given by:

$$R = \sqrt{\frac{A}{\pi}} \qquad \text{(Eq 5-3)}$$

Cooper E-80 Railroad Loads

The loading configuration used for railroad loading is the Cooper E-80 loading, which is an 80,000-lb load uniformly applied over a rectangle with dimensions 8 ft × 20 ft. The basis of these dimensions is the width of the railroad ties (8 ft) and the spacing between the drive wheels on the locomotive. Loading is based on the axle weight exerted on the track by two locomotives and their tenders coupled together in a doubleheader fashion (see Table 5-4 for loading). Railroad companies normally require encasement of pressure pipes installed under mainline track crossings. This is primarily a safety concern to prevent washout of tracks in the event of leakage.

SURCHARGE LOADS

Surcharge loads, such as footings, foundations, or other stationary loads, create pressure in the soil beneath the loaded area. This pressure will be distributed through the soil such that there is a reduction in soil pressure acting on the pipe with an increase in depth or horizontal distance from the surcharged area. (With the addition of an impact factor, the method, which follows, may be used to find the load under a heavy vehicle where the tire pattern, also called the tire imprint area, is known.)

The following method for finding the pressure acting on the pipe at a given depth directly below a surface load is based on the Boussinesq equation[9]. (Refer to Figure 5-2.) The point pressure is found by dividing the rectangular surcharge area (*ABCD*) into four subarea rectangles (*a, b, c, d*) that have a common corner, *E,* in the surcharge area and over the pipe. The surcharge load is the sum of the four subarea loads at the subsurface point. Each subarea load is calculated by multiplying an influence factor, *I,* from Table 5-5, by the surcharge pressure:

$$P_{ES} = P_a + P_b + P_c + P_d \qquad \text{(Eq 5-4)}$$

Where:

P_{ES} = surcharge load pressure at point of pipe, psf

P_a, P_b, P_c, P_d = subarea (*a, b, c,* or *d*) surcharge load, psf, defined as follows:

$$P_{(a,\,b,\,c,\,d)} = I_C W_S \qquad \text{(Eq 5-5)}$$

Table 5-4 Cooper E-80 railroad loads

Soil Height Above Pipe Crown *(ft)*	W_L *(lb/in.^{2})*	Soil Height Above Pipe Crown *(ft)*	W_L *(lb/in.^{2})*
2.5	19.3	7	11.9
3	18.4	8	10.6
4	16.8	9	9.5
5	15.1	10	8.4
6	13.4	20	3.3

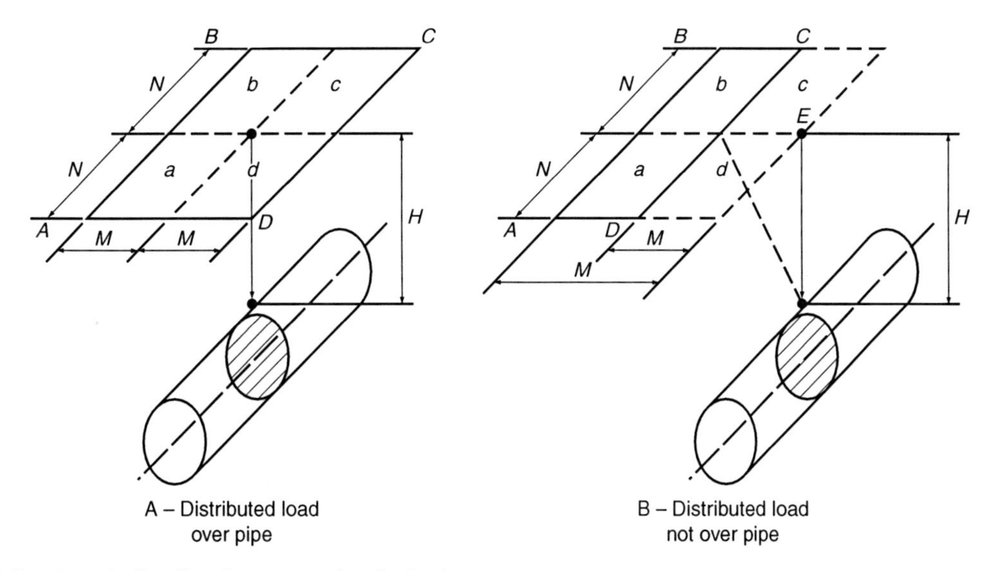

Figure 5-2 Load distribution over buried pipe

Where:

I_C = influence coefficient from Table 5-5

W_s = distributed surcharge pressure acting over ground surface, psf

If the four subareas are equal, then Eq 5-4 may be simplified to:

$$P_{ES} = 4\, I_C\, W_S \qquad \text{(Eq 5-6)}$$

The influence coefficient is dependent on the dimensions of the rectangular area and on the depth to the pipe crown. Table 5-5 influence coefficient terms depicted in Figure 5-2 are defined as:

H = vertical distance from surface to the pipe crown, ft

M = horizontal distance, normal to the pipe centerline, from the center of the load to the load edge, ft

N = horizontal distance, parallel to the pipe centerline, from the center of the load to the load edge, ft

Table 5-5 Influence coefficients

	N/H						
M/H	0.1	0.2	0.3	0.4	0.5	0.6	0.7
0.1	0.005	0.009	0.013	0.017	0.020	0.022	0.024
0.2	0.009	0.018	0.026	0.033	0.039	0.043	0.047
0.3	0.013	0.026	0.037	0.047	0.056	0.063	0.069
0.4	0.017	0.033	0.047	0.060	0.071	0.080	0.087
0.5	0.020	0.039	0.056	0.071	0.084	0.095	0.103
0.6	0.022	0.043	0.063	0.080	0.095	0.107	0.117
0.7	0.024	0.047	0.069	0.087	0.103	0.117	0.128
0.8	0.026	0.050	0.073	0.093	0.110	0.125	0.137
0.9	0.027	0.053	0.077	0.098	0.116	0.131	0.144
1.0	0.028	0.055	0.079	0.101	0.120	0.136	0.149
1.2	0.029	0.057	0.083	0.106	0.126	0.143	0.157
1.5	0.030	0.060	0.086	0.110	0.131	0.149	0.164
2.0	0.031	0.061	0.089	0.113	0.135	0.153	0.169

	N/H						
M/H	0.8	0.9	1.0	1.2	1.5	2.0	∞
0.1	0.026	0.027	0.028	0.029	0.030	0.031	0.032
0.2	0.050	0.053	0.055	0.057	0.060	0.061	0.062
0.3	0.073	0.077	0.079	0.083	0.086	0.089	0.090
0.4	0.093	0.098	0.101	0.106	0.110	0.113	0.115
0.5	0.110	0.116	0.120	0.126	0.131	0.135	0.137
0.6	0.125	0.131	0.136	0.143	0.149	0.153	0.156
0.7	0.137	0.144	0.149	0.157	0.164	0.169	0.172
0.8	0.146	0.154	0.160	0.168	0.176	0.181	0.185
0.9	0.154	0.162	0.168	0.178	0.186	0.192	0.196
1.0	0.160	0.168	0.175	0.185	0.194	0.200	0.205
1.2	0.168	0.178	0.185	0.196	0.205	0.209	0.212
1.5	0.176	0.186	0.194	0.205	0.211	0.216	0.223
2.0	0.180	0.192	0.200	0.209	0.216	0.232	0.240
∞	0.185	0.196	0.205	0.212	0.223	0.240	0.250

Interpolation may be used to find values not shown. The influence coefficient gives the portion (or influence) of the load that reaches a given depth beneath the corner of the loaded area.

A surcharge load not located directly over a pipe may still exert pressure on the pipe. Where this case occurs, the above method can be used to find the pressure on the pipe by substituting an imaginary load that covers the pipe and then subtracting that load from the overall load on the pipe.

Refer to Figure 5-2B. Because there is no surcharge directly above the pipe centerline, an imaginary surcharge load of the same pressure per unit area as the actual load is applied to the subareas c and d. The surcharge loads for subareas $a + d$ and $b + c$ are determined, then the surcharge loads from the imaginary areas c and d are deducted to find the surcharge load on the pipe.

$$P_{ES} = P_{a+d} + P_{b+c} - P_c - P_d \qquad \text{(Eq 5-7)}$$

Where terms are as previously defined above and

P_{a+d} = surcharge load combined subareas a and d, psf

P_{b+c} = surcharge load combined subareas b and c, psf

RING DEFLECTION

The following sections show how to determine the pipe's reaction to the external load. This section can be used to determine if the ring deflection in the pipe caused by an applied load is acceptable. The Wall Buckling section in this chapter examines the pipe's resistance to buckling under an applied load; and the Wall Compressive Stress section also in this chapter discusses the ring compressive stress occurring caused by an external load. For a conservative design, the pipe must be within its safe allowable limit for each of these three reactions—deflection, buckling, and ring compression.

Ring deflection is an essential response of flexible pipes to soil load. Deflection promotes arching, allows the pipe to shed load, and develops supporting reactions in the surrounding soil. Spangler introduced the Iowa Formula in 1941 to characterize the deflection's response of flexible pipe to earth pressures. In the fifties, Watkins introduced the Modified Iowa Formula, which has since been the most common method for calculating flexible pipe deflection. The Modified Iowa Formula adapted for DR is given as Eq 5-8:

$$\%\frac{\Delta Y}{D_M} = \frac{K(T_L P_E + P_L + P_{ES})}{\dfrac{2E}{3(DR-1)^3} + 0.061E'}(100) \qquad \text{(Eq 5-8)}$$

Where:

$\%\frac{\Delta Y}{D_M}$ = percent deflection

ΔY = deflection or change in diameter, in.

D_M = mean diameter, in. $(D_o - t)$

K = bedding constant, typically 0.1

T_L = time-lag factor

P_E = earth load pressure at pipe crown, psi

P_L = live load pressure at pipe crown, psi

P_{ES} = surcharge pressure at pipe crown,* psi

E = apparent modulus of elasticity of pipe material, psi (Table 5-6)

DR = dimension ratio (D_o/t)

E' = design modulus of soil reaction, psi

* P_{ES} values in psf must be converted to psi by dividing the psf value by 144.

Table 5-6 Apparent modulus of elasticity @ 73°F

Load Duration	PE 3408	PE 2406
Short term	110,000 psi	88,000 psi
Long term	28,250 psi	22,600 psi

Strictly speaking, the deflection determined in Eq 5-8 is the horizontal deflection. Most PE pressure pipes tend to deflect into a slightly elliptical shape. However, for the purpose of these calculations, the horizontal and vertical deflections can be considered equal.

Time-Lag Factor, T_L

As backfill is placed over a flexible pipe, it deflects. The deflection permits arching to occur and the redistribution of load to the sidefill soil. Because arching occurs, the pipe never has the full weight of the backfill above it, but it may have some increase in load with time. This occurs as the soil in the backfill settles and works its way downward. In addition, deflection may increase with time because of consolidation of the embedment material and the in-situ soil caused by the horizontal earth pressure at the pipe's springline. Spangler accounted for increasing deflection occurring over time by multiplying the initial deflection by the time-lag factor. Spangler proposed that the time-lag factor ranges from 1.25 to 1.5. Howard[10,11] has shown that the time-lag factor varies with the type of embedment and the degree of compaction. For design purposes, a time-lag factor of 1.5 is generally conservative for flexible pipes including PE pipe.

Modulus of Elasticity of Pipe Material, E

Janson recommends the use of the short-term pipe, elastic-modulus value in Spangler's equation where the pipe is surrounded by "frictional" soils (i.e., granular soils) and where the deflection does not continue after the first settling and soil stabilization[12]. The concept is that soil settlement around the buried pipe occurs in dynamic, discrete events as soil grains shift or fracture. Once movement occurs, soil arching redistributes the load, and no further deflection occurs for that event. Because these load increments are felt like impulse loads, the pipe resists them with its short-term elastic modulus.

Using the short-term modulus is satisfactory for most applications for PE pressure pipes installed in granular or compacted fills. However, for pipes in nonpressure applications, particularly pipes with low SDRs such as SDR 9 and SDR 11, the pipe may carry a significant portion of the earth load. As these pipes undergo stress relaxation, the apparent modulus decreases and more deflection occurs. Where deflection continues to increase with time, Janson recommends using a long-term value for modulus[13]. The long-term modulus should be considered where the pipe's stiffness term in Spangler's equation is a major portion of the pipe's resistance to deflection. For nonpressure applications, a rule-of-thumb is to use the long-term modulus, if the first term in the denominator of Eq 5-9 (pipe's stiffness term) is greater than 25 percent of the second term in the denominator (soil stiffness term).

Modulus of Soil Reaction, E'

The modulus of soil reaction represents the support stiffness of the soil surrounding the pipe in reaction to lateral pipe deflection under load. Table 5-7 gives E' values for

Table 5-7 Bureau of Reclamation values for E', modulus of soil reaction

Soil Type-Pipe Bedding Material (Unified Classification System)*	E' for Degree of Bedding Compaction, $lb/in.^2$			
	Dumped	Slight, <85% Proctor, <40% relative density	Moderate, 85%–95% Proctor, 40%–70% relative density	High, >95% Proctor, >70% relative density
Fine-grained soils (LL > 50)† Soils with medium to high plasticity CH, MH, CH-MH	No data available: consult a soil engineer, or use $E' = 0$.			
Fine-grained soils (LL < 50) Soils with medium to no plasticity CL, ML, ML-CL, with less than 25% coarse grained particles	50	200	400	1,500
Fine-grained soils (LL < 50) Soils with medium to no plasticity CL, ML, ML-CL, with more than 25% coarse grained particles	150	400	1,000	2,500
Coarse-grained soils with fines GM, GC, SM, SC‡ contains more than 12% fines				
Coarse-grained soils with little or no fines GW, GP, SW, SP‡ contains less than 12% fines	200	700	2,000	3,000
Crushed rock	1,000	1,000	3,000	3,000
Accuracy in terms of percentage deflection§	±2%	±2%	±1%	±0.5%

* ASTM D-2487, USBR Designation E-3

† LL = liquid limit

‡ Or any borderline soil beginning with one of these symbols (i.e., GM-GC, GC-SC).

§ For ±1% accuracy and predicted deflection of 3%, actual deflection would be between 2% and 4%.

NOTE: Values applicable only for fills less than 50 ft (15 m). Table does not include any safety factor. For use in predicting initial deflections only; appropriate deflection lag factor must be applied for long-term deflections. If bedding falls on the borderline between two compaction categories, select lower E' value or average the two values. Percentage proctor based on laboratory maximum dry density from test standards using 12,500 ft-lb/ft^3 (598,000 J/m^3) (ASTM D-698, AASHTO T-99, USBR Designation E-11). 1 psi = 6.9 KPa.

embedment soils from extensive field measurements taken by A. Howard[9] for the Bureau of Reclamation. The values in the table are obtained from initial average deflections. To obtain maximum predicted deflections, variability in the values in the table can be accommodated by reducing the E' value by 25 percent or by adding to the calculated deflection the percentages given at the bottom of Table 5-7.

The value of E' increases with increasing depth of cover. This is because of the increased confinement of the embedment soil by the surrounding soil. The increase in confinement stiffens the embedment soil and raises its E'. Shallow trenches produce less confinement on embedment than deeper trenches, so shallow installations typically have lower E' values. Table 5-8 gives E' values for shallow cover depths as determined by Duncan and Hartley[14]. It is suggested that for a given application, the designer compare E' values in both Table 5-7 and Table 5-8 and use the most conservative.

Table 5-8 Duncan-Hartley's values of E', modulus of soil reaction

Type of Soil	Depth of Cover, *ft*	E' for Standard AASHTO Relative Compaction, $lb/in.^2$			
		85%	90%	95%	100%
Fine-grained soils with less than 25% sand content	0–5	500	700	1,000	1,500
	5–10	600	1,000	1,400	2,000
	10–15	700	1,200	1,600	2,300
	15–20	800	1,300	1,800	2,600
Coarse-grained soils with fines (SM, SC)	0–5	600	1,000	1,200	1,900
	5–10	900	1,400	1,800	2,700
	10–15	1,000	1,500	2,100	3,200
	15–20	1,100	1,600	2,400	3,700
Coarse-grained soils with little or no fines (SP, SW, GP, GW)	0–5	700	1,000	1,600	2,500
	5–10	1,000	1,500	2,200	3,300
	10–15	1,050	1,600	2,400	3,600
	15–20	1,100	1,700	2,500	3,800

The E' value in Table 5-7 can be used to a depth of 50 ft. At deeper depths, the E' value should be corrected to account for its increase with depth of cover, or alternate deflection calculations, such as Burns and Richards[3], should be performed.

For pipes installed in trenches, the support stiffness developed depends on the combined stiffness of the embedment material, immediately adjacent to the pipe, plus the native soil in the trench wall. Thus E' is found by combining the modulus of soil reaction of the embedment soil, E'_E with the modulus of soil reaction of the native soil, E'_N. For pipes installed in embankments, E' equals E'_E. For the native soils, typical values of the soil reaction modulus, E'_N are given in Table 5-9 in terms of presumptive soil values. (See Howard[15].)

The modulus of soil reaction, E', to be used for design of pipe in a trench is obtained from the following equation:

$$E' = S_C E'_E \quad \text{(Eq 5-9)}$$

Where:

E' = design modulus of soil reaction, psi

S_C = soil support factor from Table 5-10

E'_E = modulus of soil reaction of embedment, psi

Deflection Limits

The limiting deflection (in percent) is established in consideration of the geometric stability of the deflected pipe, hydraulic capacity, and the maximum fiber strain occurring in the pipe wall. Generally, thermoplastic pipes remain geometrically stable up to deflections of 25 to 30 percent. In addition, Jansen observed that for PE, pressure-rated pipe, subjected to soil pressure only, "no upper limit from a practical design point of view seems to exist for the bending strain"[13]. Thus, for nonpressure applications a

Table 5-9 Values of E'_N, modulus of soil reaction for native soil, from Howard[14]

Native In Situ Soils				
Granular		Cohesive		
Std. Penetration ASTM D1586, *blows/ft*	Description	Unconfined Compressive Strength (TSF)	Description	E'_N *(psi)*
>0–1	very, very loose	>0–0.125	very, very soft	50
1–2	very loose	0.125–0.25	very soft	200
2–4	very loose	0.25–0.50	soft	700
4–8	loose	0.50–1.00	medium	1,500
8–15	slightly compact	1.00–2.00	stiff	3,000
15–30	compact	2.00–4.00	very stiff	5,000
30–50	dense	4.00–6.00	hard	10,000
>50	very dense	>6.00	very hard	20,000
Rock	—	—	—	≥50,000

Table 5-10 Soil support factor, S_c

E'_N/E'_E*	B_d/D_o 1.5	B_d/D_o 2.0	B_d/D_o 2.5	B_d/D_o 3	B_d/D_o 4	B_d/D_o 5
0.1	0.15	0.30	0.60	0.80	0.90	1.00
0.2	0.30	0.45	0.70	0.85	0.92	1.00
0.4	0.50	0.60	0.80	0.90	0.95	1.00
0.6	0.70	0.80	0.90	0.95	1.00	1.00
0.8	0.85	0.90	0.95	0.98	1.00	1.00
1.0	1.00	1.00	1.00	1.00	1.00	1.00
1.5	1.30	1.15	1.10	1.05	1.00	1.00
2.0	1.50	1.30	1.15	1.10	1.05	1.00
3.0	1.75	1.45	1.30	1.20	1.08	1.00
5.0	2.00	1.60	1.40	1.25	1.10	1.00

* E'_N equals the modulus of soil reaction for the native soil; E'_E equals the modulus of soil reaction for the embedment soil; B_d equals the trench width at the pipe springline (in.); and D_o equals the pipe outside diameter (in.).

7.5 percent deflection limit provides a large safety factor against instability and is considered a safe design deflection. Pressurized buried pipes have fiber tensile strain from both bending deformation and internal pressure and so may have slightly higher wall strains than nonpressure pipes. This effect is offset in part by the tendency of the internal pressure to reround the pipe. Because of the combined strain, deflection limits for pressurized pipe are related to the pipe's dimension ratio (*DR*). Design deflection for pressurized pipe can be found in Table 5-11. Pipe standards for PE pressure pipe do not require postinstallation deflection inspection.

WALL BUCKLING

When buried pipes are subjected to external loads such as negative internal pressure (partial vacuum), groundwater, or extremely high earth loads, an instability can occur

Table 5-11 Design deflection for pressure pipe

DR	Design Deflection as % of Diameter
32.5	7.5
26	7.5
21	7.5
17	6.0
13.5	6.0
11	5.0
9	4.0
7.3	3.0

* Based on long-term design deflection of buried pressurized pipe given in ASTM F1962.

in the pipe wall that may lead to large inward deformations called buckling. A pipe's resistance to buckling is increased by the constraining effect of the surrounding soil. To develop this support, the pipe must have a cover of at least four feet or one pipe diameter (whichever is greater). The following equation for constrained buckling can be used to calculate a buried pipe's resistance to external pressure (loads)[16]:

$$P_{CA} = \frac{5.65}{N}\sqrt{R_b B' E' \frac{E}{12(DR-1)^3}} \quad \text{(Eq 5-10)}$$

Where:

P_{CA} = allowable external pressure for constrained pipe, psi
N = safety factor, typically 2.0 for polyethylene pipe
R_b = buoyancy reduction factor
B' = soil elastic support factor
E' = modulus of soil reaction, psi
E = apparent modulus of elasticity (Table 5-6)
DR = dimension ratio (D_o/t)

$$R_b = 1 - 0.33\frac{H_W}{H} \quad \text{(Eq 5-11)}$$

H_W = groundwater height above pipe, ft
H = depth of cover, ft

$$B' = \frac{1}{1 + 4_e^{-0.065H}} \quad \text{(Eq 5-12)}$$

e = natural log base number, 2.71828

The allowable external pressure should be greater than the summation of external loads applied to the pipe:

$$P_{CA} > P_E + P_L + P_{ES} + P_V \qquad \text{(Eq 5-13)}$$

Where:

P_E, P_L, P_{ES} = external pressure at pipe crown caused by earth load, live load, and surcharge load*, respectively, psi

P_V = internal vacuum in pipe, psi

* P_{ES} values in psf must be converted to psi by dividing the psf value by 144.

In addition to verifying the constrained buckling resistance of a pipe, if the pipe is subject to internal vacuum, the pipe should be capable of withstanding the vacuum load independent of the soil support. The following equation may be used to determine the allowable negative pressure or external pressure an unconstrained (or unsupported) pipe can resist without collapsing:

$$P_{UA} = \left(\frac{2E}{1-\mu^2}\right)\left(\frac{1}{DR-1}\right)^3\left(\frac{f_o}{N}\right) \qquad \text{(Eq 5-14)}$$

Where:

P_{UA} = allowable external pressure for unconstrained pipe, psi

E = apparent modulus of elasticity, psi (Table 5-6)

μ = Poisson's ratio

long-term loading—0.45

short-term loading—0.35

DR = dimension ratio (D_o/t)

f_o = ovality compensation factor (see Figure 5-3)

N = safety factor, generally 2.0

It should be noted that the modulus of elasticity and Poisson's ratio are a function of the duration of the anticipated load.

When the pipe is buried less than 4 ft deep or less than a full diameter, the constraining effect of the soil may not fully develop. In this case, engineering judgment should be used to determine whether or not Eq 5-10 is suitable for determining the pipe's buckling resistance to live load. Alternatively, the buckling resistance may be calculated using Eq 5-14.

WALL COMPRESSIVE STRESS

The earth pressure applied to a buried pipe creates a compressive thrust stress in the pipe wall. When pipelines are pressurized, the compressive stress is usually canceled by the tensile thrust stresses from pressurization. Buried pressure lines are subject to a net external pressure only when shut down or when experiencing a vacuum. During these events, the compressive stress is generally not present long enough to be significant,

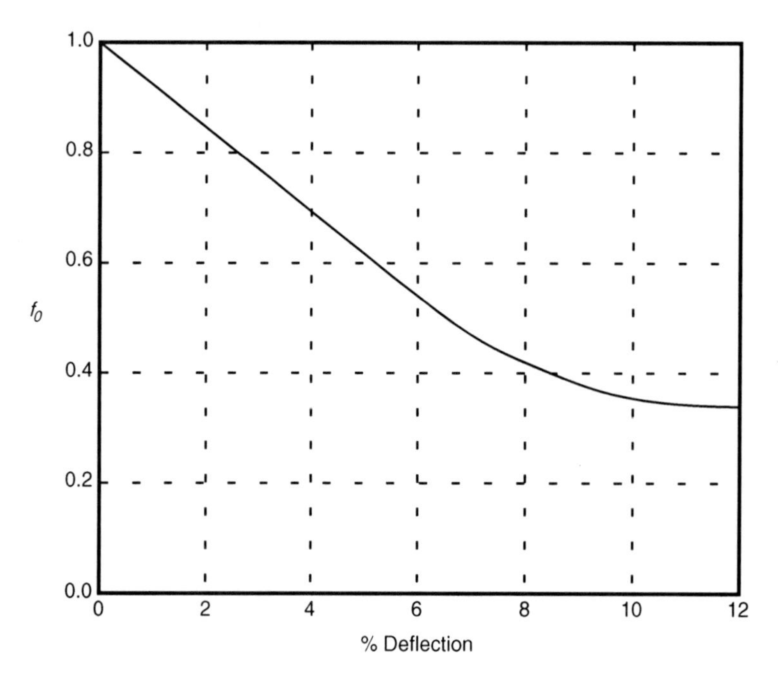

Figure 5-3 Ovality correction factor

because the short-term design stress of PE is considerably higher than the long-term design stress. (Buckling must be considered for all events applying external pressures. See the Wall Buckling section.) Pipes with large depths of cover and operating at low pressures may have compressive stresses in the pipe wall. The following equation can be used to determine the compressive stress:

$$S = \frac{P_E(D_o - t)}{2t} = \frac{P_E(DR - 1)}{2} \quad \text{(Eq 5-15)}$$

Where:

S = hoop compressive wall stress, psi

P_E = earth pressure on pipe, psi

D_o = pipe outside diameter, in.

t = wall thickness, in.

The compressive wall stress should be kept less than the allowable compressive stress of the material. A conservative approach is to assume that the sustained allowable long-term compressive stress is equal to 1.25 times the allowable hydrostatic design stress (HDS) adjusted for temperature. (HDS is the HDB times the design factor (*DF*), which is 0.5 for water applications.) Accordingly, in the case of PE3408, the allowable long-term compressive stress at 73.4°F (23°C) is 1,000 psi, and for PE2406, the allowable long-term compressive stress at 73.4°F (23°C) is 800 psi.

DESIGN WINDOW

Polyethylene pipes having *DRs* equal to or less than 21 and installed per the following conditions meet the design deflection limits of the Ring Deflection section, have a safety factor against buckling of at least 2.0 per Eq 5-10, and have a compressive stress not exceeding 1,000 psi per Eq 5-15. Therefore, for pipes within this design window no calculation is needed. See Example No. 1.

Design Window

1. Pipe has a *DR* no greater than *DR* 21.
2. Minimum depth of cover is 2 ft (0.61 m), except when subject to AASHTO H2O truck loadings; in which case, the cover is equal to the greater of 3 ft (0.9 m) or one pipe diameter.
3. There is essentially no dead surface load imposed over the pipe and no groundwater above the surface. Provisions for flotation of shallow cover pipe have been provided.
4. Maximum height of cover is 25 ft (7.62 m).
5. The native soil has a modulus of soil reaction of at least 1,000 psi and the unit weight does not exceed 120 pcf (18.87 kN/m^3). (See Table 5-9.)
6. The embedment materials are coarse-grained, stable, and have been compacted to at least 85 percent Standard Proctor Density.
7. The pipe has been installed so that stress generated by pipe deformation and secondary loads are limited by
   a. adherence to recommendations for minimum bending radius, and
   b. installation of the pipe in accordance with the recommendations given in Chapter 8.

EXAMPLE NO. 1

Calculate deflection, allowable buckling pressure, and compressive stress for a DR21 HDPE water pipe buried 25 ft deep in a clayey-sand soil. The native soil contains about 15 percent clay and has an E'_N of 1,000 psi. The native soil is used for embedment, which is compacted to 85 percent of Standard Proctor Density. At the time of installation, the groundwater table is below the invert so that the embedment can be placed and compacted. After construction, the groundwater reaches the surface. Assume the conditions of the design window section are met. Pipe dimensions are given in Table 3-1 and 3-2. The window is independent of diameter and pipe dimensions are not required for this calculation.

Cover depth (H) = 25 ft

Groundwater height above pipe (H_W) = 25 ft

Bedding coefficient (K) = 0.1

Modulus of soil reaction (E'_N) for native soil = 1,000 psi

Modulus of soil reaction (E') for the embedment = 1,000 psi

Soil saturated unit weight (w) = 120 pcf

Time-lag factor (T_L) = 1.5

The potential maximum earth pressure on the pipe is:

$$P_E = wH = 120(25) = 3{,}000 \text{ psf}$$

To determine maximum earth pressure in psi, divide by 144:

$$P_E = \frac{3{,}000}{144} = 20.8 \text{ psi}$$

At 25 ft of cover, H2O load pressure, P_L, is negligible. No surcharge pressure, P_{ES}, is present. The apparent modulus of elasticity of the HDPE pipe material is 110,000 psi. The modulus of soil reaction, E', is equal to 1,000 psi, because both E'_E and E'_N equal 1,000 psi. Spangler's Iowa Formula is:

$$\%\frac{\Delta Y}{D_M} = \frac{K(T_L P_E + P_L + P_{ES})}{\frac{2E}{3(DR-1)^3} + 0.061E'}(100)$$

Substituting values into Spangler's Iowa Formula yields:

$$\%\frac{\Delta Y}{D_M} = \frac{0.1(1.5(20.8 \text{ psi}) + 0 + 0)}{\frac{2(110{,}000 \text{ psi})}{3(21-1)^3} + 0.061(1{,}000 \text{ psi})}(100)$$

$$\%\frac{\Delta Y}{D_M} = \frac{3.12 \text{ psi}}{9.17 \text{ psi} + 61 \text{ psi}}(100) = 4.4\%$$

The anticipated design deflection, 4.4 percent, is less than the allowable design deflection, 7.5 percent, given in Table 5-11 for DR21 pipe.

Next, determine the maximum allowable buckling pressure, P_{CA}, for the pipe. Because groundwater will be at the surface, $H' = 25$ and the long-term apparent modulus of HDPE is 28,250 psi:

$$R = 1 - 0.33(25/25) = 0.67$$

$$B' = \frac{1}{1 + 4_e^{-0.065(25)}} = 0.56$$

$$P_{CA} = \frac{5.65}{2}\sqrt{(0.67)0.56(1{,}000 \text{ psi})\frac{28{,}250 \text{ psi}}{12(21-1)^3}} = 29.7 \text{ psi}$$

The allowable buckling pressure, 29.7 psi, which includes a 2:1 safety factor, is greater than the applied earth pressure of 20.8 psi.

The compressive stress in the DR21 pipe is given by Eq 5-15, which can be written in terms of the pipe *DR* as:

$$S = \frac{P_E(DR - 1)}{2}$$

Substituting in values for 25 ft of cover.

$$S = \frac{20.8\ \text{psi}(21 - 1)}{2} = 208\ \text{psi}$$

The compressive stress is less than the allowable compressive stress of 1,000 psi.

EXAMPLE NO. 2

What will be the deflection of a *DR* 17 HDPE water pipe buried in a trench? If an embedment material of sandy clay with less than 50 percent fines is assumed, the modulus of soil reaction, E'_E, is 400 psi when installed with slight compaction. Also assume the following:

Pipe nominal diameter = 8 in.

Pipe mean diameter (D_M) = 8.625 in. – 0.507 in. = 8.12 in.

Trench width at pipe springline (B_d) = 20 in.

Cover depth (H) = 12 ft

Bedding coefficient (K) = 0.1

Modulus of soil reaction (E'_N) for native soil = 700 psi

Native soil unit weight (w) = 120 pcf

Time-lag factor (T_L) = 1.0

Internal operating pressure = 100 psi

Ignore the rerounding effects of the internal pressure. To calculate the design modulus of soil reaction, E', first determine the following ratios:

$$\frac{E'_N}{E'_E} = \frac{700}{400} = 1.75$$

$$\frac{B_d}{D_M} = \frac{20}{8.12} = 2.46$$

Next, obtain the soil support factor, S_C, by extrapolation in Table 5-10.

$$S_C = 1.12$$

$$E' = S_C E'_E = 1.12(400) = 448 \text{ psi}$$

$$P_E = \frac{wH}{144} = \frac{120(12)}{144} = 10 \text{ psi}$$

There is no live load pressure, P_L, or surcharge pressure, P_{ES}, and the apparent modulus of elasticity of the HDPE pipe material is 110,000 psi.

$$\%\frac{\Delta Y}{D_M} = \frac{K(T_L P_E + P_L + P_{ES})}{\frac{2E}{3(DR-1)^3} + 0.061E'}(100)$$

Substituting values into Spangler's Iowa Formula yields:

$$\%\frac{\Delta Y}{D_M} = \frac{0.1(1.0(10 \text{ psi}) + 0 + 0)}{\frac{2(110{,}000 \text{ psi})}{3(17-1)^3} + 0.061(448 \text{ psi})}(100)$$

$$\%\frac{\Delta Y}{D_M} = \frac{1.0 \text{ psi}}{17.9 \text{ psi} + 27.3 \text{ psi}}(100) = 2.2\%$$

EXAMPLE NO. 3

Assume that the pipe in Example No. 1 is buried only 3 ft deep under an asphalt paved road. Further assume the same native soil as in Example No. 1 but assume that the pipe embedment material is a coarse-grained soil compacted to 90 percent Standard Proctor Density. The E'_E for this material is 2,000 psi.

$$\frac{E'_N}{E'_E} = \frac{700}{2{,}000} = 0.35$$

$$\frac{B_d}{D_M} = \frac{20}{8.12} = 2.46$$

Extrapolating from Table 5-10 to obtain the soil support factor, S_C.

$$S_C = 0.78$$

$$E' = S_C E'_E = 0.78\ (2{,}000) = 1{,}560 \text{ psi}$$

$$P_E = \frac{wH}{144} = \frac{120(3)}{144} = 2.5 \text{ psi}$$

Table 5-3 gives the live load for an H20 vehicle at 3 ft as 5.4 psi, and Table 5-1 gives impact factor at 3 ft of 1.4.

There is no surcharge pressure, P_{ES}, and the apparent modulus of elasticity of the HDPE pipe material is 110,000 psi.

$$\%\frac{\Delta Y}{D_M} = \frac{K(T_L P_E + P_L + P_{ES})}{\frac{2E}{3(DR-1)^3} + 0.061E'}(100)$$

Substituting values into Spangler's Iowa Formula yields:

$$\%\frac{\Delta Y}{D_M} = \frac{0.1(1.0(2.5 \text{ psi}) + 1.4(5.4 \text{ psi}) + 0)}{\frac{2(110{,}000 \text{ psi})}{3(17-1)^3} + 0.061(1{,}560 \text{ psi})}(100)$$

$$\%\frac{\Delta Y}{D_M} = \frac{1.0 \text{ psi}}{17.9 \text{ psi} + 95.2 \text{ psi}}(100) = 0.9\%$$

EXAMPLE NO. 4

Assume that the DR17 HDPE pipe in Example No. 1 is subjected to a full vacuum during a water hammer event with the groundwater table level at the ground surface. Assume that the saturated unit weight of the soil is 130 pcf. Verify that the combined pressure caused by the vacuum, groundwater, and overburden soil is less than the allowable buckling pressure of the pipe with a 2 to 1 safety factor.

During a vacuum event, the total pressure applied to the pipe will equal

$$P_V = 14.7 \text{ psi} + \frac{130 \text{ pcf}(12 \text{ ft})}{144} = 25.5 \text{ psi}$$

Find the allowable buckling pressure, P_{CA}, for the pipe. Because groundwater is at the surface $H' = 12$ and

$$R = 1 - 0.33(12/12) = 0.67$$

$$B' = \frac{1}{1 + 4_e^{-0.065(12)}} = 0.35$$

$$P_{CA} = \frac{5.65}{2}\sqrt{(0.67)0.35(448\ \text{psi})\frac{110{,}000\ \text{psi}}{12(17-1)^3}} = 43.3\ \text{psi}$$

Compare the allowable buckling pressure with the applied external pressure:

$$P_{CA} > P_V$$

EXAMPLE NO. 5

When vacuum can occur in the pipe, it is usually prudent to check whether the pipe can withstand the vacuum, P_V, without soil support (if the depth is shallower than anticipated or pipe is uncovered). Check the above pipe to see if it can withstand the internal vacuum without soil support. Assume 2 percent ovality.

$$P_{UA} = \frac{2(110{,}000\ \text{psi})}{1-(0.35)^2}\left(\frac{1}{17-1}\right)^3\frac{0.84}{2.0} = 25.7\ \text{psi}$$

$$P_{UA} > V = 14.7\ \text{psi}$$

REFERENCES

1. Petroff, L.J. 1990. "Stress Relaxation Characteristics of the HDPE Pipe-Soil System," Proceedings of Conference on Pipeline Design and Installation, ASCE, Las Vegas, NV.
2. Spangler, M.G. and Handy, R.L. 1982. *Soil Engineering,* Harper Row Publishers, New York, 4th Ed.
3. Burns, J.Q. and Richards, R.M. 1964 "Attenuation of Stresses for Buried Cylinders," Proceedings of Symposium of Soil Structure Interaction, University of Arizona.
4. Moore, I.D. 1987. *The Elastic Stability of Shallow Buried Tubes,* Geotechnique 37, No. 2, 151-161.
5. Watkins, R.K. and Anderson, L.R. 2000. *Structural Mechanics of Buried Pipes.* CRC Press, Washington, DC.
6. American Iron and Steel Institute, 1994. *Handbook of Steel Drainage and Highway Construction Products,* American Iron and Steel Institute, Washington, DC, 4th Ed.
7. Stewart, H.E. and O'Rourke, T.D. 1991. *Live Loads for Pipeline Design at Railroads and Highways,* Proc. Pipeline Crossings, ASCE, Denver, CO.
8. Holtz, R.D. and Kovacs, W.D. 1981. *An Introduction to Geotechnical Engineering,* Prentice-Hall, Englewood Cliffs, New Jersey.
9. Oey, H.S., Greggerson, V.L., and Womack, D.P. 1984 "Buried Gas Pipelines Under Vehicular Crossings," ASCE, *Journal of Transportation Engrng.* Vol. 110, No. 2.
10. Howard, A. "Modulus of Soil Reaction (E') Values for Buried Flexible Pipe," *Journal of Geotechnical Engineering Division, ASCE* Vol. 103, No. Gt.
11. Howard, A.K. 1981. "The USBR Equation for Predicting Flexible Pipe Deflection," Proc. International Conf. On Underground Plastic Pipe, ASCE, New Orleans.
12. Janson, L.E. 1985. "Investigation of the Long Term Creep Modulus for Buried Polyethylene Pipes Subjected to Constant Deflection." Proc. International Conference on Advances in Underground Pipeline Engrg., ASCE, Madison, WI.
13. Janson, L.E. 1991. *Long-Term Studies of PVC and PE Pipes Subjected to Forced Constant Deflection,* Report No. 3, KP-Council, Stockholm, Sweden.

14. Hartley, J. and Duncan, J.M. 1982. *Evaluation of the Modulus of Soil Reaction, E′, and Its Variation with Depth,* Report No. UCB/GT/82-02, University of California, Berkeley.
15. Howard, A.K. 1996. *Pipeline Installation.* Johnson Printing, Boulder, Colorado.
16. Cagle, L. et al. 1982. "Recommendations for Elastic Buckling Design Requirements for Buried Flexible Pipe," Proc., Part 1, AWWA 1982 Annual Conference, "Better Water for the Americas."

This page intentionally blank.

Chapter 6

Joining and Fittings

HEAT FUSION JOINING

An integral part of any pipe system is the method used to join the system components. Proper engineering design of a system will consider the type and effectiveness of the techniques used to join the piping components and appurtenances as well as the durability of the resulting joints. The integrity and versatility of the joining techniques used for ANSI/AWWA C901[1] and C906[2] PE pipe allow the designer to take advantage of the performance benefits of PE in a wide variety of applications. ANSI/AWWA C901 and C906 PE piping products are joined using heat fusion and mechanical methods, such as flanges and compression couplings. Joining and connection methods will vary depending on requirements for internal pressure and restraint against longitudinal movement (thrust load capacity).

When heat fusion joining, it is essential that the heat fusion joining procedures used are qualified by making heat fusion joints with the procedures, and then conducting tests on the joints to demonstrate joint integrity. Furthermore, persons making field joints should have experience in making heat fusion joints and should be qualified by appropriate training that includes making joints using the procedures that are then tested to demonstrate joint integrity. Pipe and joining equipment manufacturers should be consulted for information about joining procedures, training, and appropriate tests for demonstrating joint integrity.

Joint design limitations and manufacturer's joining procedures must be observed; otherwise, the joint or products adjacent to the joint may leak or fail, which may result in property damage or hazards to persons. Use the tools and components required to construct and install joints in accordance with manufacturer's recommendations. Field joints are controlled by and are the responsibility of the field installer.

General Procedures

PE pressure pipe and fittings are joined to each other and to other piping components by heat fusion and mechanical connection methods that seal and provide restraint against pullout. The preferred method for joining PE pipe and fittings is by heat fusion. Properly made heat fusion joints are fully restrained and do not leak. PE is mainly joined to other materials by mechanical means such as compression fittings, flanges, mechanical joint adapters, and transition fittings. Mechanical joints may require an additional restraint device to prevent pullout movement. Joining methods that do not provide restraint against pullout are not acceptable for use in PE pressure piping systems. Fittings of various types are available for a variety of close quarter or repair applications.

Heat Fusion Joining

There are two methods for producing heat fusion joints: conventional heat fusion, where heat is applied with an external heating plate, and electrofusion, where an electric heating element is an integral part of the electrofusion fitting. Conventional heat fusion includes butt, saddle or sidewall, and socket fusion. Electrofusion is used to produce socket and saddle heat fusion joints.

The principle of heat fusion is to heat and melt the two joint surfaces and force the melted surfaces together, which causes the materials to mix and fuse into a monolithic joint. The applied fusion pressure is maintained until the joint has cooled. When fused according to the pipe and/or fitting manufacturers' procedures, the joint becomes as strong as or stronger than the pipe itself in both tensile and pressure properties. The sections that follow describe general guidelines for each of these heat fusion methods.

The manufacturer's recommended heat fusion procedures should always be used.

Butt Fusion

The most widely used method for joining individual lengths of PE pipe is by the butt fusion of the pipe ends as illustrated in Figure 6-1. This technique, which precludes the need for specially modified pipe ends or couplings, produces a permanent, economical, and flow-efficient joint. Field-site butt fusion may be made readily by trained operators using high quality butt fusion machines (Figure 6-2) that secure and precisely align the pipe ends for the fusion process. Butt fusion joining is applicable to pipes that have the same nominal outside diameter and wall thickness (*DR*). Although special clamp adapters for making angled (mitered) fusion joints are available from some equipment suppliers, field miter butt fusion joining is never used in PE pressure piping systems because the pressure capacity of a miter butt fusion joint made using pipe having the same *DR* as the main pipe is lower than the pressure capacity of the main pipe. When direction changes are required, fittings that have pressure capabilities equal to or greater than the pressure capability of the main pipe are required. See the Mainline Fittings section in this chapter.

Butt fusion involves

1. Securely fastening the components to be joined.
2. Facing the component ends being joined.
3. Aligning the component ends.
4. Melting the component ends.
5. Applying force to fuse and join the component ends together.
6. Maintaining applied force until the joint has cooled.

Always observe manufacturer's recommended fusion procedures. Manufacturer's recommended procedures should be consulted for specific information such as joining to other brand, wall thickness variation, and alignment. Modern hydraulic butt fusion machines can be equipped with a data logging device to record critical joining parameters, including heater surface temperature and heating, fusion, and cooling pressures and times. The specifications engineer or purchaser of a piping system may specify a comparison of the recorded parameters to qualified parameters to confirm fusion quality.

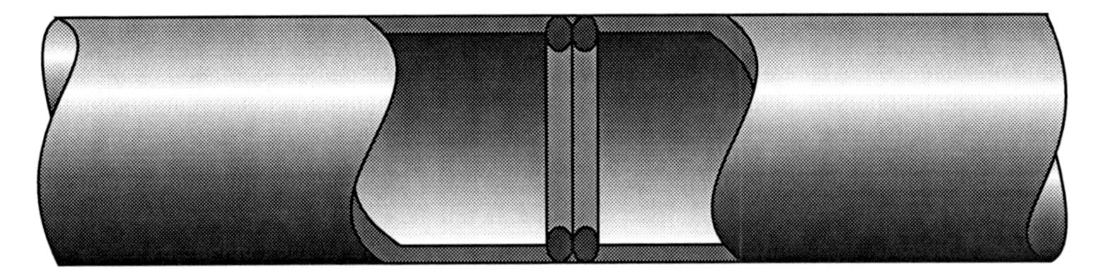

Figure 6-1 Butt fusion joint

Figure 6-2 Typical butt fusion machine (butt fusion machines are available to fuse pipe up to 65 in. in diameter)

Data from the data logging device can be transferred to a personal computer for permanent storage. The data logging device can also be used as a training aide for operators. For information about butt fusion joining in addition to manufacturer's recommended procedures, see PPI Technical Report TR-33[3] and ASTM D2657[4]. The generic butt fusion joining procedures published in PPI TR-33 have been evaluated by major North American PE pipe manufacturers and have been shown by testing to produce sound joints when used with their products and when used to join to the other manufacturer's products. A listing of manufacturers endorsing PPI TR-33 generic joining procedures is published in PPI TR-33.

All sizes of ANSI/AWWA C901 and C906 pipe and butt fusion fittings may be field joined by butt fusion. Butt fusion equipment ranges from manual units to hydraulically assisted, semiautomatic machines for pipes up to 65 in. Butt fusion equipment can be fitted with inserts and will accommodate a range of pipe sizes.

For estimating purposes, Table 6-1 presents the approximate number of field joints for a typical construction day. Actual joining rates may be different, depending on size and wall thickness, jobsite conditions, product staging, equipment condition, crew size and experience, and handling equipment. Table 6-1 rates do not include fusion machine setup time, and do not apply to tie-in joints or to butt fusion in the trench.

Table 6-1 Approximate joining rates for butt fusion

Pipe Size IPS or DIPS *(in.)*	Approximate Joining Rate Butt Fusions per 8–10 hr Day
4–8	48–24
10–18	24–12
20–24	16–10
36–48	10–6
65	8–4

It is frequently easier and faster to bring lightweight PE pipe lengths to the machine, rather than the machine to the pipe. Electric heating irons, especially large ones, take some time to heat up, so when the machine is relocated, the heater must be brought back up to temperature before joining, and the machine must be set up for fusion. For fusion in the trench, setup may involve removing the fusion clamp assembly from the cart, and additional excavation may be necessary to clear open clamps, facing and heating tools, and to allow machine removal. In the trench, it is recommended to remove the machine from the pipe, rather than lift the pipe out of the machine. This involves rotating the machine around the pipe, then lifting it off, or dropping the machine down below the pipe and moving it to the side to clear the pipe. Additional excavation from machine removal clearance may be required.

Setup time is minimized when pipe lengths are fed through the machine and joined into long strings. Common construction practice is to set up a *fusing station* to join lengths into long strings. The strings are positioned along the pipe run and joined together by moving the fusion machine from string to string. At the fusing station, pipe lengths are stockpiled near the fusion machine. The first two lengths are joined and pulled through the machine so that a third length can be joined to the second, and so forth. Typical strings are 500 ft to 1,500 ft or longer. Joining strings is also called *tie-in joining*.

Bead Removal

Butt fusion produces a double-roll melt bead on the inside and the outside of the pipe. Internal bead removal is not necessary because its effect on hydraulic flow is negligible.

Saddle Fusion

Conventional saddle fusion. The technique to join a saddle to a pipe, illustrated in Figure 6-3, consists of simultaneously heating both the external surface of the pipe and the matching surface of the saddle-type fitting with concave- and convex-shaped heating tools until both surfaces reach proper fusion temperature. A saddle fusion machine that has been designed for this purpose is required. A bolster plate that helps round and support the pipe is frequently used when saddle fusing to coiled pipe. A bolster plate may still be necessary for saddle fusion even where rerounding and straightening equipment that corrects the ovality and curvature of 2-in. IPS through 6-in. IPS coiled pipes has been used during pipe installation. Observe the saddle fusion equipment manufacturer's operating instructions and ensure that all appropriate equipment necessary for making the saddle fusion on the main pipe is employed.

Saddle fusion joining involves

1. Installing the saddle fusion machine on the main pipe.

Figure 6-3 Conventional saddle fusion joint

Figure 6-4 Typical electrofusion saddle fusion

2. Abrading the main pipe surface where the saddle will be applied.
3. Abrading the mating saddle fitting surface.
4. Simultaneously melting the mating pipe main and saddle surfaces.
5. Removing the heater and applying force to join and fuse the melted surfaces together.
6. Maintaining the applied force until the joint has cooled.

Always observe the manufacturer's recommended fusion procedures. For information about saddle fusion joining in addition to manufacturer's recommended procedures, see PPI TR-41[5] and ASTM D2657.

Electrofusion saddle fusion. The main distinction between conventional saddle fusion and electrofusion is the method by which the heat is applied to the surface of the pipe and the fitting. Electrofusion joining is accomplished by application of electrical energy to a wire coil or other electrically conductive material that is an integral part of the fitting. Unlike conventional saddle fusion fittings, electrofusion fittings are clamped or otherwise affixed to the pipe before heat energy is applied. The electrical energy passing through the wire coil of the fitting creates resistance heat, expanding both the pipe and the fitting at the interface, which creates the pressure required to mix the materials for joining. Figure 6-4 illustrates typical electrofusion saddle fusion joining.

Electrofusion saddle fusion involves:

1. Preparing the pipe surface (scraping and cleaning).
2. Aligning the fitting to the pipe and securing the fittings with a holding clamp.
3. Attaching control box leads and starting the automatic electrofusion cycle.
4. Allowing the joint to cool.

Always observe manufacturer's recommended fusion procedures. For information about electrofusion joining in addition to manufacturer's recommended procedures, see ASTM F1290[6].

Socket Fusion

Conventional socket fusion. This technique consists of simultaneously heating and melting the external surface of a pipe end and the internal surface of a socket fitting, inserting the pipe end into the socket, and holding it in place until the joint cools. Figure 6-5 illustrates typical socket fusion joining. Mechanical equipment is typically used to hold the pipe and the fitting, especially for 2-in. and larger sizes, and to assist in alignment and joining.

Conventional socket fusion may be used with 4-in. and smaller pipe and fittings. Socket fusion heater faces are manufactured to ASTM F1056[7]. Field socket fusion tools are hand-held, and for larger sizes, two persons are usually needed to make joints.

Socket fusion involves

1. Squaring and preparing the pipe end.
2. Using a depth gauge to fit a cold ring clamp to the pipe at the correct insertion depth.
3. Simultaneously heating and melting the pipe end and the fitting socket.
4. Joining the components together.
5. Holding the components together until the joint has cooled.

Always observe manufacturer's recommended fusion procedures. For information about socket fusion joining in addition to manufacturer's recommended procedures, see ASTM D2657.

Electrofusion socket fusion. The major distinction between conventional socket fusion and electrofusion socket fusion is the method by which the heat is applied to the surface of the pipe and the fitting. Electrofusion fittings have a heating element built into the fitting. Electrofusion joining is accomplished by applying electrical energy to the heating element through a control box. Unlike conventional socket fusion fittings, electrofusion fittings are clamped or otherwise affixed to the pipe before heat energy is applied. The electrical energy passing through the integral heating element creates heat, which melts and expands, and then heats and melts the mating pipe surface. Melt expansion at the interface also creates the pressure required to mix and fuse the materials together. Figure 6-6 illustrates a typical electrofusion socket joint, and Figure 6-7 illustrates an electrofusion control box and fitting.

Electrofusion joining involves

1. Preparing the pipe surface (scraping and cleaning).
2. Aligning the fitting to the pipe and (if necessary) securing the fitting with a holding clamp.
3. Attaching control box leads and starting the automatic electrofusion cycle.
4. Allowing the joint to cool.

Electrofusion is the only heat fusion procedure that does not require longitudinal movement of one of the joining components. It is frequently used where both pipes are constrained, such as for repairs or tie-in joints in the trench.

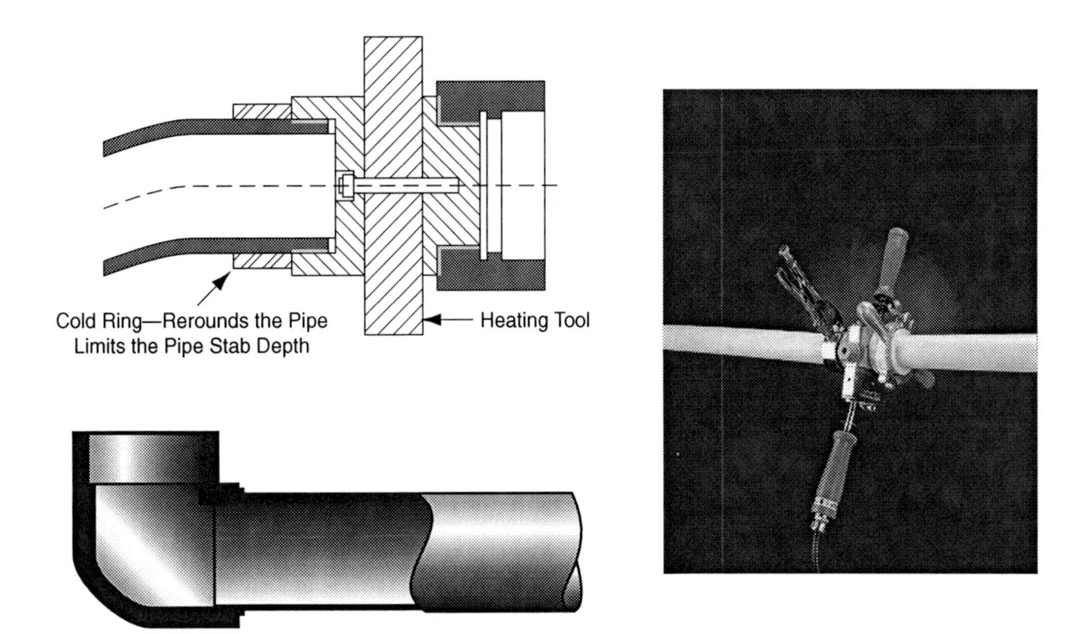

Figure 6-5 Socket fusion joining

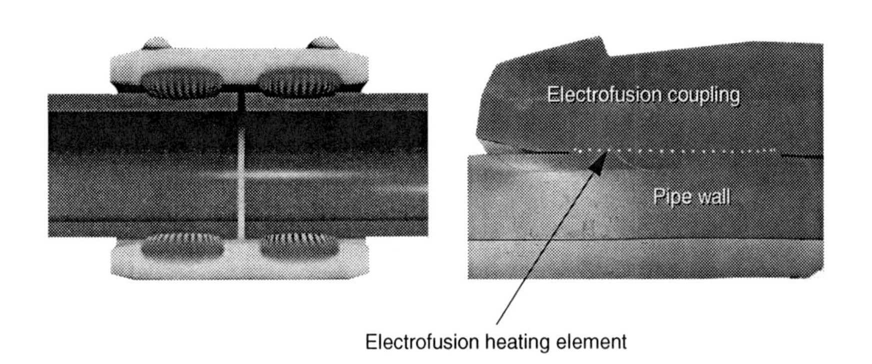

Figure 6-6 Typical electrofusion joint

Figure 6-7 Typical electrofusion fitting and control box (lower right)

Always observe manufacturer's recommended fusion procedures. For information about electrofusion joining in addition to manufacturer's recommended procedures, see ASTM F1290.

MECHANICAL JOINING

ANSI/AWWA C901 and C906 pipe and fittings may be joined to themselves and to other piping materials using various mechanical joining methods including mechanical compression couplings, flanges, mechanical joint (MJ) adapters, and transition fittings. All mechanical joint products used with ANSI/AWWA C901 and C906 pressure piping systems must provide restraint against pullout. Joining devices and components with joints that seal but do not restrain must be provided with additional external mechanical restraint. Seal-only mechanical joining devices without restraint are not suitable for joining PE pressure piping. (**NOTE:** Thrust blocks do not provide restraint against pullout and are not a substitute for external mechanical restraint.)

Mechanical Compression Fittings

Mechanical compression fittings typically have a body that is a pressure-containing component that fits over the outside diameter of the pipe, a threaded compression nut or follower gland with bolting arrangement, and elastomeric seal rings or gaskets. Fittings that provide resistance to pipe pullout will also have some type of gripping device to prevent the pipe from pulling out of the fitting. All mechanical compression fittings apply a compressive load on the pipe to create a pressure seal or activate the restraint device. Most manufacturers of these couplings recommend the use of insert stiffeners to reinforce the pipe against OD compression from the coupling. Manufacturer's recommended procedures should be consulted for specific information. The pipe stiffener is positioned in the pipe bore so that its location is under the gasket and the gripping device (if applicable) of the fitting. The stiffener keeps the pipe from collapsing from the compression loads, which could result in leakage or even failure of the restraining feature if the fitting provides restraint. If an additional external restraining device is also required, the stiffener should extend under the restraint clamps. When the compression nut or bolt arrangement is tightened, the elastomeric gasket is compressed against the pipe surface to effect a pressure tight seal. Tightening can also activate a gripping device if the fitting is designed to also provide restraint (Figure 6-8).

Stab-type mechanical fittings do not require the tightening of bolts or compression nuts. When pushed onto the pipe end, the fitting supports the pipe end and seals and provides restraint, generally by a mechanism where a slight movement of the pipe activates a gripping device to prevent pullout. Stab-type fittings are generally used on smaller pipe and tubing sizes.

Insert fittings are also used to join smaller PE pipe and tubing. The fitting joint is made by inserting a barbed nipple into the pipe bore and tightening a clamp or drawing a clamp over the pipe outside diameter over the barbed nipple.

When properly installed, stab-type and insert fittings will provide a leak tight connection and some resistance to pull out. They are typically available for use on 2-in. and smaller pipe sizes.

Most mechanical fittings can be installed with basic hand tools. However, some fittings may require specialized tools for installation, which if required, are available from the fitting manufacturer.

Figure 6-8 Mechanical compression coupling with restraint—PE restrained by electrofusion flex restraints—PVC pipe restrained using a tapered gripping ring

Mechanical fittings are produced in practically all the sizes of PE pipe available. Fitting configurations may be limited in some sizes because of market demands. Fitting manufacturers should be contacted for application suggestions and product availability.

Mechanical compression joints to PE pipe are fully restrained against thrust load only if pressure and tensile loads cause the pipe to yield before the pipe and fitting disjoins. Mechanical joints that provide full thrust restraint are designed to mechanically compress the pipe OD against a rigid tube or stiffener in the pipe bore. Some mechanical compression fittings provide restraint but to a lesser extent. For these fittings, the designer should determine that the restraint provided is sufficient to resist applied thrust loads from temperature change, internal pressure, and other static and dynamic thrust loads. If the fitting's thrust restraint is inadequate, external restraint such as clamps and tie rods or in-line anchors should be installed.

Internal stiffeners are available from a number of manufacturers and may be custom sized for the pipe, or adjustable with a wedging device that when hammered into the stiffener, expands the stiffener to a tight fit in the pie bore. ID measurements taken from the actual pipe may be required for custom fit stiffeners. Caution should be used when using mechanical fittings to ensure that the stiffener OD is the correct size for the specific DR of the pipe to be joined.

PE Flange Adapter Connection

When joining to PE, metal, or other piping materials or if a pipe section capable of disassembly is required, PE flange adapters, as depicted in Figure 6-9, are available. Flange adapters and stub ends are designed so that one end is sized for butt fusion to the plastic pipe. The other end is made with a flange and a serrated sealing surface. A backup ring fits behind the flange, and when bolted to a mating flange, the backup ring compresses against the flange to provide a seal and thrust resistance. A stiffener is not required. Flange adapters have a long spigot that is clamped in the butt fusion machine. Stub ends have a short spigot and require a stub-end holder for butt fusion. Flange adapters differ from stub ends by their overall length. A flange adapter is longer allowing it to be clamped in a fusion machine like a pipe end. The backup ring is fitted to the flange adapter before fusion, so external fusion bead removal is not required.

A stub end is short and requires a special stub-end holder for butt fusion. Once butt fused to the pipe, the external bead must be removed so the backup ring can be fitted

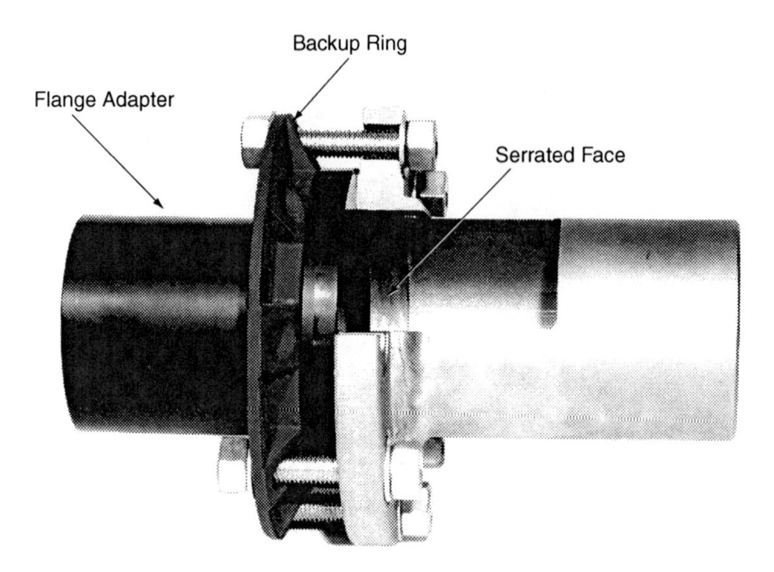

Figure 6-9 PE flange adapter

behind the sealing surface flange. In the field, flange adapters are usually preferred over stub ends.

Backup rings of various pressure ratings are available. The manufacturer should be consulted to ensure that the backup rings specified are correctly pressure rated for the application. ANSI Class 125 or ANSI/AWWA C207[8] Class D backup rings are typically used for 160 psi and lower pressure ratings and ANSI Class 150 for higher pressure ratings. Backup ring materials are steel, primer-coated steel, epoxy-coated steel, or stainless steel. Ductile iron and fiberglass are also available. In belowground service, coatings and cathodic protection may be appropriate to protect metal backup rings from corrosion. One edge of the backup ring bore must be radiused or chamfered. This edge fits against the back of the flange adapter sealing surface flange. An all-PE flange without a backup ring is not recommended because PE flanges require uniform pressure over the entire sealing surface. Without a backup ring, a PE flange will leak between the bolts.

When used with backup rings having sufficient pressure rating, PE flange connections are pressure rated and have the same surge allowance as pipe having the same DR and made from the same material.

Flange adapter connection involves

1. Fitting the backup ring against the flange adapter sealing flange and clamping the flange adapter in the butt fusion machine.
2. Butt fusing the flange adapter to the PE pipe.
3. Aligning the sealing surface of the flange adapter to the mating flange, installing the gasket (if required), and installing the flange bolts. Mating flange alignment *before* tightening is critical in obtaining a satisfactory flange joint and is particularly critical for large flange connections. Never use the flange bolts to draw the two sections of pipe into alignment. Tightening misaligned flanges can cause flange failure. Prior to fit-up, lubricate flange bolt threads, washers, and nuts with a nonfluid lubricant. Gasket and flange sealing surfaces must be clean and free of cuts or gouges. Fit the flange components together loosely.

4. Hand-tighten bolts and recheck alignment. Adjust alignment if necessary. All flange bolts should be tightened to the same torque value by turning the nut. Tighten each bolt according to the patterns and torques recommended by the flange manufacturer. Tightening the flange bolts in an indexed, top-bottom, left-right crossing pattern; that is, tighten the top bolt, then the bottom bolt 180° across from it, then move 90° to the left and tighten that bolt, then 180° to the right across from it and tighten that bolt. Index the crossing pattern one bolt to the right and repeat with the next set of four flange bolts. Flange bolts must be evenly torqued in stages to provide proper sealing. Tighten all flange bolts in pattern sequence at one torque level before increasing to the next torque level. Tightening in this pattern sequence ensures that the sealing surfaces are uniformly compressed for a proper seal. Depending on bolt size, final torque is typically 100 to 150 ft-lb, and at least three torque increments are used. A torque wrench should be used for tightening. PE and the gasket (if used) will undergo some compression set. Therefore, retightening is recommended about an hour or so after all flange bolts are torqued to the final torque value the first time. In pattern sequence, retighten each bolt to the final torque value. For high pressure or environmentally sensitive or critical pipelines, a third tightening, about four hours after the second, is recommended.

Flange gasketing. At lower pressure, typically 80 psi or less, a gasket is usually not required. At greater pressure, the serrated surface of the flange adapter helps hold the gasket in place. Gaskets may be needed for connections between PE and non-polyethylene flanges. If used, gasket materials should be chemically and thermally compatible with the internal fluid and the external environment and should be of appropriate hardness, thickness, and style. Elevated temperature applications may require higher temperature capability. Gasket thickness should be about $^1/_8$ in. – $^3/_{16}$ in. (3–5 mm) and about 55–75 durometer Shore D hardness. Too soft or too thick gaskets may blow out under pressure. Overly hard gaskets may not seal. Common gasket styles are full face or drop-in. Full-face style gaskets are usually applied to larger sizes, because flange bolts hold a flexible gasket in place while fitting the components together. Drop-in style gaskets are usually applied to smaller pipe sizes.

Flange bolting. Mating flanges are usually joined together with hex bolts and hex nuts, or threaded studs and hex nuts. Bolting materials should have tensile strength equivalent to at least SAE Grade 3 for pressure pipe service. Corrosion resistant materials should be considered for underground, underwater, or other corrosive environments. Flange bolts are sized $^1/_8$ in. smaller than the blot hole diameter. Flat washers should be used between the nut and the backup ring. Flange bolts must span the entire width of the flange joint and provide sufficient thread length to fully engage the nut.

Flange assembly. Surface or abovegrade flanges must be properly supported to avoid bending stresses. Belowgrade flange connections to heavy appurtenances, such as valves or hydrants, or to metal pipes, require a support foundation of compacted, stable granular soil (crushed stone), or compacted cement stabilized granular backfill, or reinforced concrete. Flange connections adjacent to pipes passing through structural walls must be structurally supported to avoid shear loads.

Special cases. When flanging to brittle materials, such as cast iron, accurate alignment and careful tightening are necessary. Tightening torque increments should not exceed 10 ft-lb. PE flange adapters and stub ends are not full-face, so tightening places a bending stress across the flange face. Over-tightening, misalignment, or uneven tightening can break brittle material flanges.

When joining a PE flange adapter or stub end to a flanged butterfly valve, the inside diameter of the pipe flange should be checked for valve disk rotation clearance. The open valve disk may extend into the pipe flange. Valve operation may be restricted if the pipe flange interferes with the disk. If disk rotation clearance is a problem, a tubular spacer may be installed between the mating flanges, or custom fabricated flange adapters that are modified for butterfly valve disk rotation should be used. Flange bolt length must be increased by the length of the spacer.

Mechanical Flange Adapters

Mechanical flange adapters are also available as shown in Figure 6-10. This fitting combines a mechanical compression coupling on one end with a flange connection on the other to provide a connection between flange fittings and valves to plain end pipes. A stiffener is required in the bore of the PE pipe. Mechanical flange adapters may or may not provide restraint against pipe pullout as part of the design, and an alternative means of restraint should be used where the mechanical flange adapter does not provide restraint. Contact the fittings manufacturer for assistance in selecting the appropriate style for the application.

Mechanical Joint Adapters

Mechanical joint (MJ) adapters are used to transition from a PE piping system to a bell-and-spigot ductile iron or PVC piping system. See Figure 6-11. The MJ adapter is designed for butt fusion or electrofusion to a PE pipe and connection to a standard ANSI/AWWA C153[9] or C110[10] ductile iron mechanical joint fitting, valve, or pipe. The mechanical joint end of the MJ adapter is fitted with a standard mechanical joint gasket, which is seated against a circumferential rib and extends into the mechanical joint hub. A standard mechanical joint gland fits against the back of the rib and is bolted to the mechanical joint hub flange. Bolts longer than supplied in a standard gland pack are required. MJ adapters are normally supplied with the gasket, bolts, nuts, and gland ring. MJ adapters are available in size 3-in. IPS through 54-in. IPS.

The MJ adapter connection is a fully self-restrained joint and does not require additional restraint.

Factory Assembled Transition Fittings

Factory assembled transition fittings allow the user to make a connecting joint in field application between dissimilar piping materials. Department of Transportation (DOT) Category 1 transition fittings seal tight with pressure and are fully restrained.

Transition fittings are supplied with a PE pipe for heat fusion and a metal end for connection by welding, taper pipe threads, flange, or shouldered for mechanical connection (Figure 6-12). The metal end is typically coated steel but may also be available in other metals, such as brass or stainless steel. Contact the transition-fitting manufacturer for product availability information.

MAINLINE FITTINGS

For the mainline installation, fittings are available for changes in direction or changes in size. The flexibility of PE allows for gradual redirection of the pipe, often eliminating the need for 22-½°and lesser angle bends. Molded butt fusion fittings are available in PE pipe and tubing sizes ½-in. CTS through 12-in. IPS. Larger sizes are available as a "fabricated" fitting. The term *fabricated* means that the part is made by joining sections of pipe or molded fittings to form the desired fitting configuration (Figure 6-13).

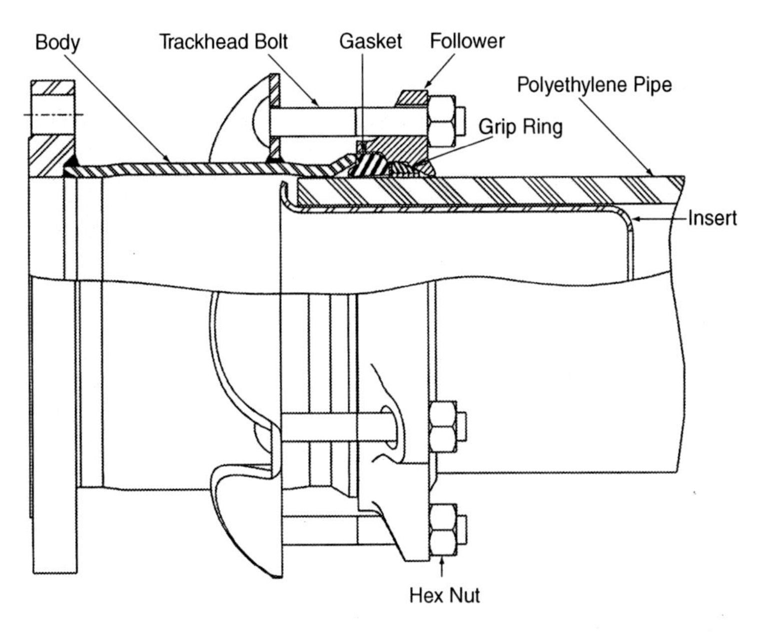

Figure 6-10 Mechanical flange adapter

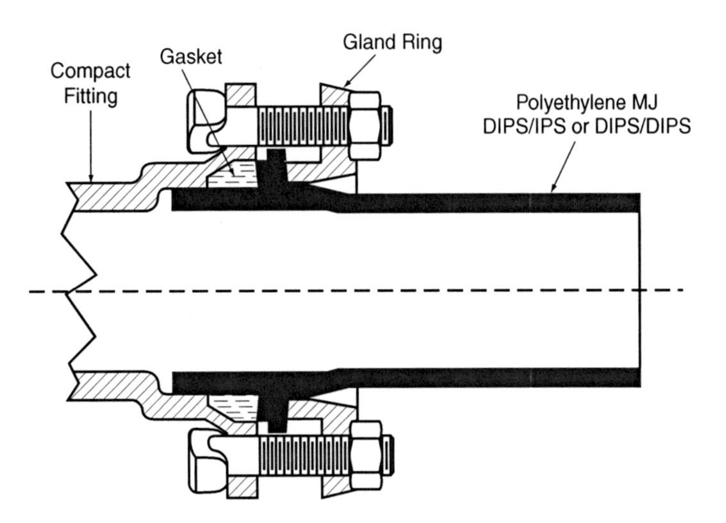

Figure 6-11 MJ adapter

Molded and fabricated butt fittings are available in most styles including inline reducer couplings, tees, reducing tees, end caps, elbows, and crosses in both size-to-size and reducing sizes. With the manufacturing option of "fabrication," almost any configuration of butt fitting can be produced. Molded butt fittings should comply with ASTM D3261[11]. Fabricated fittings should comply with ANSI/AWWA C906 and ASTM F2206[12] to ensure full pressure rating. (**NOTE:** Field fabricated elbows do not comply with ANSI/AWWA C906 or ASTM F2206 and are not permissible in pressure service.)

A completely heat-fused PE system is fully restrained and does not require additional mechanical restraints. Larger fabricated fittings (16 in. and above) are typically installed at the end of a pipe section so that only one outlet is connected before lowering into the trench. The remaining connections are made in the trench using butt fusion, electrofusion, or mechanical connections. Attempting to move and lower a fabricated fitting that is connected to multiple pipe sections frequently damages the fitting.

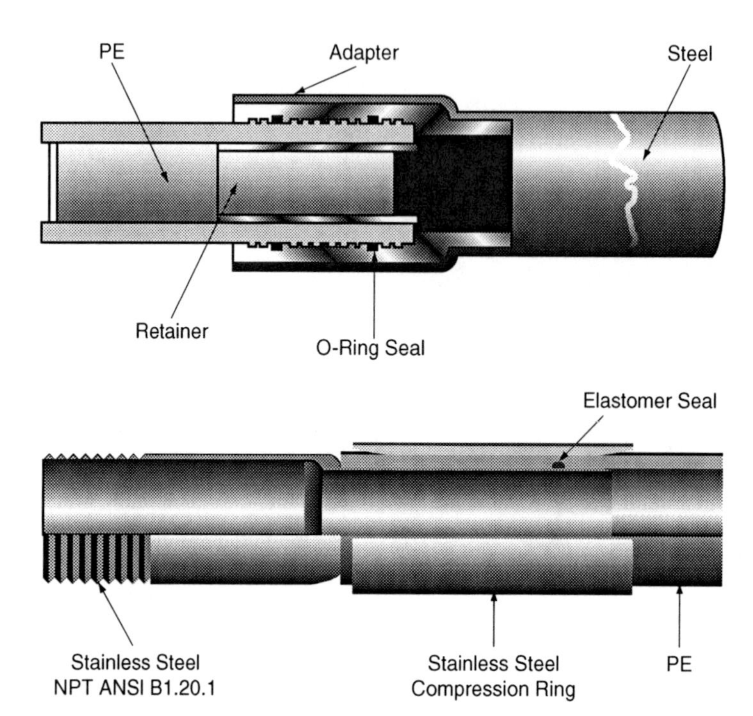

Figure 6-12 Transition fittings

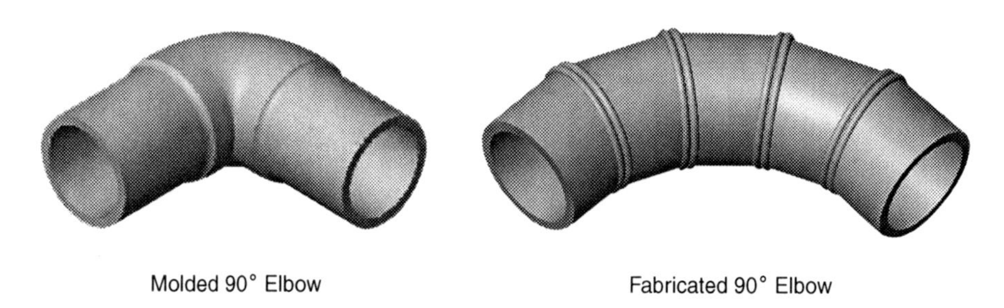

Figure 6-13 Molded and fabricated elbows

BRANCHING AND TAPPING

During pipe laying, a lateral connection may be provided by installing an inline equal or reducing tee in the main line. During or after the main is laid, service saddles, branch saddles, and tapping tees may be installed using sidewall fusion, electrofusion, or mechanical means.

Service saddles and branch saddles may be tapped with standard tapping tools and procedures for pressure and nonpressure tapping. The branch saddle is connected to the ductile gate valve using MJ adapters or flange adapters. The distance from the tangent of the main to the outside flange of the gate valve should not exceed the travel distance of the tapping tool. The cutter should be designed for use with PE pipe, with few teeth and large chip clearance (Figures 6-14 and 6-15).

For larger branch connection, mechanical tapping saddles specifically designed for use with PE may be used (Figure 6-16). Always inspect the surface of the pipe under the saddle to ensure that it is free of scratches and surface damage that might allow a leak path.

(NOTE: Blocking is removed during embedment.)

Figure 6-14 Electrofusion branch saddle connected to gate valve by MJ adapters

Figure 6-15 Conventional saddle fusion branch saddles

SERVICE CONNECTIONS

Saddle tapping tees have internal cutters and do not require external valving. The service line should be connected before tapping.

Saddle Tapping Tees

Saddle tapping tees are typically used for service line connections to provide a means to tap a live main without special tapping equipment. Tapping tees include an internal "punch" or cutter as part of the tapping tee assembly, for perforating the main. Punches for the tapping operation range in size from $^1/_4$ in. to 1 in. and will produce a very clean hole of the same size as the punch cutter while retaining the "coupon" cut from the pipe wall. Saddle bases are contoured for sidewall fusion or electrofusion to mains from $1^1/_4$-in. IPS through 8-in. IPS (Figure 6-17).

For larger diameter, higher pressure pipes with thick walls, it is necessary to ensure that the punch is long enough to punch through the pipe wall. Outlets of tapping tees are commonly available in butt or socket style in sizes from $^1/_2$-in. CTS through $1^1/_4$-in. IPS.

Figure 6-16 Mechanical tapping saddle

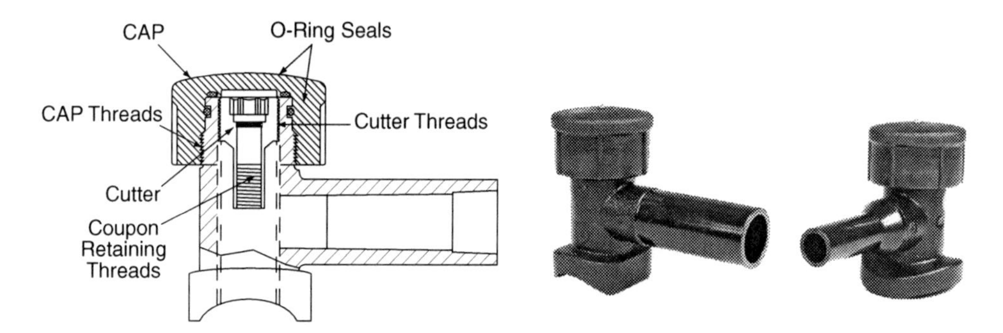

Figure 6-17 Saddle tapping tees

High Volume Saddle Tapping Tees (HVTT)

High volume saddle tapping tees (HVTT) are a larger version of the saddle tapping tee described previously. The term *high volume* indicates a larger size that may be used for 1¼-in. IPS or 2-in. IPS main extensions, branches to a main, or connecting a service line. Punch sizes range from 1¼-in. to 1⅞-in. The outlets are typically available for butt fusion 1¼-in. IPS and 2-in. IPS sizes. Saddle bases are contoured for saddle fusion or electrofusion to main sizes from 2-in. IPS through 8-in. IPS (Figure 6-18).

Corporation stop saddles are available in both electrofusion and mechanical styles. These saddles are threaded for a corporation stop and live tapping tools for tapping of the mainline. A tapping tool cutter should be designed for use with PE pipe with few teeth and large chip clearance. Bolt-on style saddles should be designed for the movement of PE pipe. Always inspect the surface of the pipe under the saddle to be free of scratches that might allow a leak path (Figure 6-19).

REFERENCES

1. ANSI/AWWA C901, *Polyethylene (PE) Pressure Pipe and Tubing, ½ In. (13 mm) Through 3 In. (76 mm), For Water Service,* American Water Works Association, Denver, CO.
2. ANSI/AWWA C906, *"AWWA Standard for Polyethylene (PE) Pressure Pipe and Fittings, 4 In. (100 mm) Through 63 In.(1,575 mm) for Water Distribution and Transmission,"*

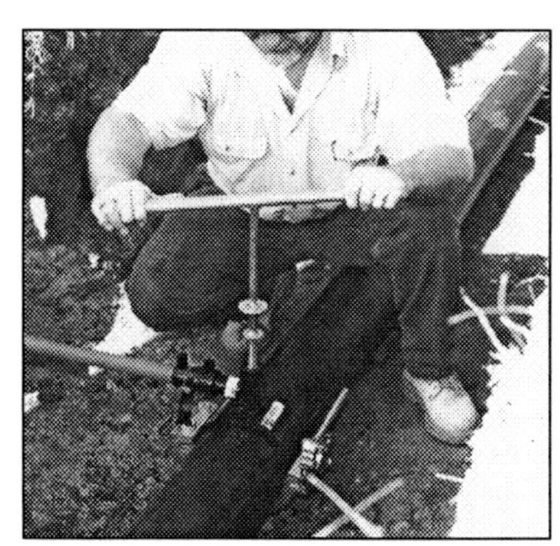

Figure 6-18 High volume tapping tees (HVTT)

Figure 6-19 Corporation stop saddles

American Water Works Association, Denver, CO.

3. PPI TR-33 *Generic Butt Fusion Joining Procedure for Polyethylene Gas Pipe.*
4. ASTM D2657 *Standard Practice for Heat Fusion Joining of Polyolefin Pipe and Fittings.*
5. PPI TR-41 *Generic Saddle Fusion Joining Procedure for Polyethylene Gas Piping.* Plastic Pipe Institute, Washington, DC.
6. ASTM F1290 *Standard Practice for Electrofusion Joining Polyolefin Pipe and Fittings.* ASTM International, West Conshohoken, PA.
7. ASTM F1056 *Standard Specification for Socket Fusion Tools for Use in Socket Fusion Joining Polyethylene Pipe or Tubing and Fittings.* ASTM International, West Conshohoken, PA.
8. ANSI/AWWA C207, *Steel Pipe Flanges for Waterworks Service - Sizes 4 In. Through 144 In. (100 mm Through 3,600 mm),* American Water Works Association, Denver, CO.
9. ANSI/AWWA C153, *ANSI Standard for Ductile-Iron Compact Fittings, 3 in. (76 mm) Through 64 in. (1,600 mm), for Water Service,* American Water Works Association, Denver, CO.
10. ANSI/AWWA C110 *ANSI Standard for Ductile-Iron and Gray-Iron Fittings, 3 in. through 48 in. (76 mm through 1,219 mm), for Water,* American Water Works Association, Denver, CO.
11. ASTM D3261 *Standard Specification for Butt Heat Fusion Polyethylene (PE) plastic fittings for PE Plastic Pipe and Tubing.* ASTM International, West Conshohoken, PA.
12. ASTM F2206 *Standard Specification for Fabricated Fittings of Butt-Fused Polyethylene (PE) Plastic Pipe, Fittings, Sheet Stock, Plate Stock, or Block Stock.* ASTM International, West Conshohoken, PA.

This page intentionally blank.

Chapter 7

Transportation, Handling, and Storage of Pipe and Fittings

After the piping system has been designed and specified, the piping system components must be obtained. Typically, project management and purchasing personnel work closely together so that the necessary components are available when needed for the upcoming construction work.

RECEIVING INSPECTION

Few things are more frustrating and time consuming than not having what you need, when you need it. Before piping system installation begins, an important initial step is a receiving inspection of incoming products. Construction costs can be minimized and schedules maintained by checking incoming goods to be sure the parts received are the parts that were ordered, and that they arrived in good condition and are ready for installation.

PE pipe, fittings, and fabrications are shipped by commercial carriers who are responsible for the products from the time they leave the manufacturing plant until the receiver accepts them. Pipe and fabricated fittings and structures are usually shipped on flatbed trailers. Smaller fittings may be shipped in enclosed vans or on flatbed trailers depending on size and packaging. Molded fittings are usually boxed and shipped by commercial parcel services.

PRODUCT PACKAGING

Depending on size, PE piping is produced in coils or in straight lengths. Coils are stacked together into silo packs. Straight lengths are bundled together in bulk packs or loaded on the trailer in strip loads. Standard straight lengths for conventionally extruded pipe are 40 ft long; however, lengths up to 60 ft long may be produced. State transportation restrictions on length, height, and width usually govern allowable load configurations. Higher freight costs will apply to loads that exceed length, height, or width restrictions. Although PE pipe is lightweight, weight limitations may restrict load size for very heavy wall or longer length pipe.

ANSI/AWWA C901 PE pipe is usually produced in coils. Coil lengths will vary with the pipe size. ANSI/AWWA C906 PE pipe is usually produced in 40 ft straight lengths. Other pipe lengths are possible; however, state transportation restrictions on load length, height, weight, and width will govern allowable truckload configurations. Consult with the pipe manufacturer for pipe length and freight information.

Figures 7-1 through 7-3 are general illustrations of truckload and packaging configurations. Actual truckloads and packaging may vary from the illustrations.

Small fittings are packaged in cartons, which may be shipped individually by package carriers. Large orders may be palletized and shipped in enclosed vans. Large fittings and custom fabrications may be packed in large boxes on pallets or secured to pallets.

CHECKING THE ORDER

When a shipment is received, it should be checked to ensure that the correct products and quantities have been delivered. Several documents are used. The purchase order or the order acknowledgment lists each item by its description and the required quantity. The incoming load will be described in a packing list, which is attached to the load. The descriptions and quantities on the packing list should match those on the purchase order or the order acknowledgment.

The carrier will present a bill of lading that generally describes the load as the number of packages the carrier received from the manufacturing plant. The order acknowledgment, packing list, and bill of lading should all be in agreement. Any discrepancies must be reconciled among the shipper, the carrier, and the receiver. The receiver should have a procedure for reconciling any such discrepancies.

LOAD INSPECTION

All AWWA PE pipe is marked to identify size, dimension ratio, pressure class, material, and the manufacturer's production code. When the pipe is received, it should be visually inspected to verify that the correct product was received. The product should also be checked for damage that may have occurred during transit. Look for fractures, kinking, deep gouges, or cuts. Minor scratches or scuffing will not impair serviceability of the PE pipe and fittings. However, pipe with gouges or cuts in excess of 10 percent of the product wall should not normally be used. The length of pipe affected by the damaged section may be cut out and the remainder of the product reused. The commercial carrier and manufacturer should be advised immediately of any damage or discrepancies.

When pipe installation involves saddle fusion joining, diesel smoke on the pipe's outside surface may be a concern because it may reduce the quality of saddle fusion joints. Covering at least the first third of the load with tarpaulins effectively prevents smoke damage. If smoke tarps are required, they should be in place covering the load when it arrives.

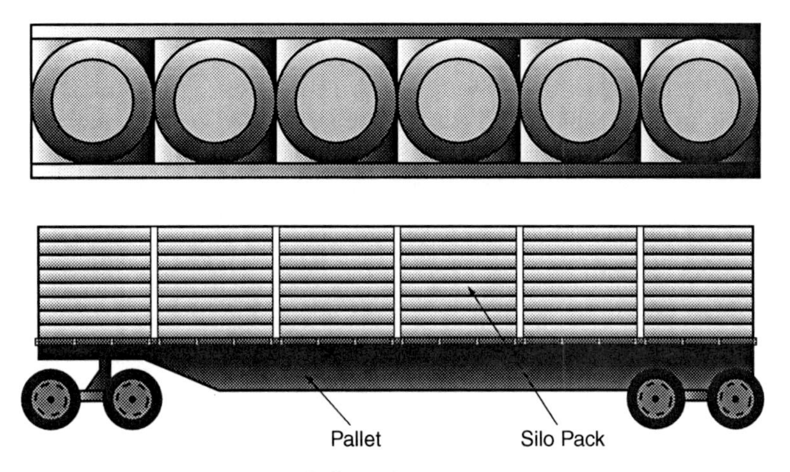

Figure 7-1 Typical silo pack truckload (40-ft trailer)

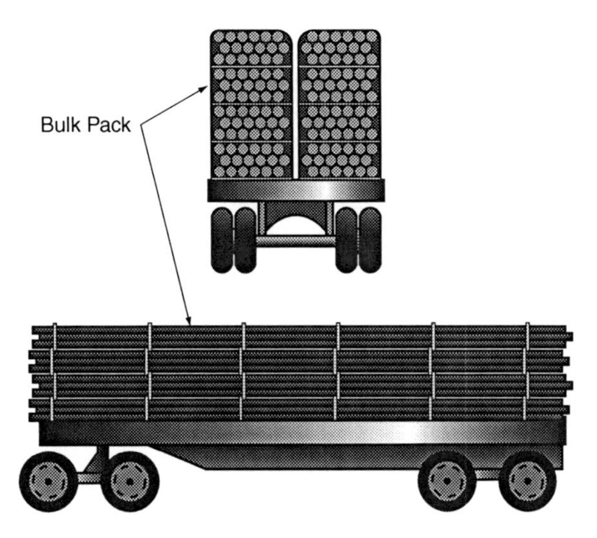

Figure 7-2 Typical bulk pack truckload (40-ft trailer)

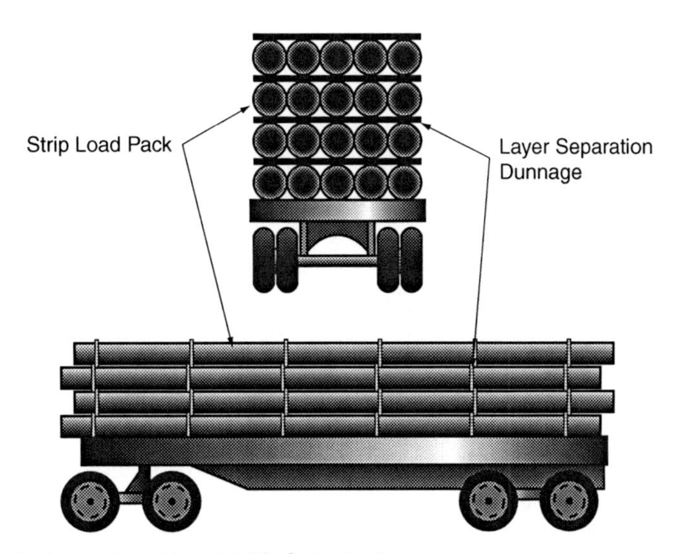

Figure 7-3 Typical strip load truckload (40-ft trailer)

RECEIVING REPORT AND REPORTING DAMAGE

The delivering truck driver will ask the person receiving the shipment to sign the bill of lading and acknowledge that the load was received in good condition. Any damage, missing packages, etc., should be noted on the bill of lading at that time.

UNLOADING INSTRUCTIONS

Manufacturer's recommended unloading instructions are given to the driver before he leaves the manufacturing plant. Unloading personnel should get these instructions from the driver and review them before unloading the truck.

Before unloading the shipment, there must be adequate, level space to unload the shipment. The truck should be on level ground with the parking brake set and the wheels chocked. Unloading equipment must be capable of safely lifting and moving pipe, fittings, fabrications, or other components.

NOTE: Unloading and handling must be performed safely. Unsafe handling can result in damage to property or equipment and be hazardous to persons in the area. Keep unnecessary persons away from the area during unloading.

NOTE: Only properly trained personnel should operate unloading equipment.

UNLOADING SITE REQUIREMENTS

The unloading site must be relatively flat and level. It must be large enough for the carrier's truck, the load handling equipment and its movement, and for temporary load storage. Silo packs and other palletized packages should be unloaded from the side with a forklift. Nonpalletized pipe, fittings, or fabrications should be unloaded from above with lifting equipment and wide web slings, or from the side with a forklift.

HANDLING EQUIPMENT

Appropriate unloading and handling equipment of adequate capacity must be used to unload the truck. Safe handling and operating procedures must be observed. Pipe, fittings, or fabrications must not be pushed or dumped off the truck or dropped.

Although PE piping components are lightweight compared to similar components made of metal, concrete, clay, or other materials, larger components can be heavy. Lifting and handling equipment must have adequate rated capacity to lift and move components from the truck to temporary storage. Equipment such as a forklift, a crane, a side boom tractor, or an extension boom crane is used for unloading.

When using a forklift or forklift attachments on equipment such as articulated loaders or bucket loaders, lifting capacity must be adequate at the load center on the forks. Forklift equipment is rated for a maximum lifting capacity at a distance from the back of the forks (Figure 7-4). If the weight-center of the load is farther out on the forks, lifting capacity is reduced.

Before lifting or transporting the load, forks should be spread as wide apart as practical, forks should extend completely under the load, and the load should be as far back on the forks as possible.

NOTE: During transport, a load on forks that are too short or too close together, or a load too far out on the forks, may become unstable and pitch forward or to the side and result in damage to the load or property or hazards to persons.

Lifting equipment, such as cranes, extension boom cranes, and side boom tractors, should be hooked to wide web choker slings that are secured around the load or to lifting lugs on the component. Only wide web slings should be used. Wire rope slings and

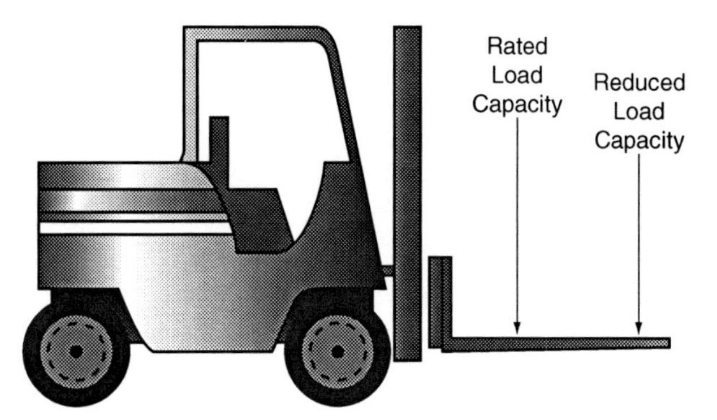

Figure 7-4 Forklift load capacity

chains can damage components and should not be used. Spreader bars should be used when lifting pipe or components longer than 20 ft.

NOTE: Before use, inspect slings and lifting equipment. Equipment with wear or damage that impairs function or load capacity should not be used.

Pipe that is packaged in coils, especially in diameters above 2 in., is difficult to handle without proper equipment. Lifting and handling equipment having sufficient capacity for the weight and size of the coil is required. Further, coiled pipe larger than 2 in. can have ovality exceeding 5 percent. Specialized equipment that straightens and rounds the pipe at the time of installation may be required.

UNLOADING LARGE FABRICATIONS

Large fabrications should be unloaded using a wide web choker sling and lifting equipment such as an extension boom crane, crane, or lifting boom. Do not use stub outs, outlets, or fittings as lifting points, and avoid placing slings where they will bear against outlets or fittings.

PREINSTALLATION STORAGE

The size and complexity of the project and the components will determine preinstallation storage requirements. For some projects, several storage or staging sites along the right-of-way may be appropriate, while a single storage location may be suitable for another job.

The site and its layout should provide protection against physical damage to components. General requirements are for the area to be of sufficient size to accommodate piping components, to allow room for handling equipment to get around them, and to have a relatively smooth, level surface free of stones, debris, or other material that could damage pipe or components, or interfere with handling. Pipe may be placed on 4-in. wide wooden dunnage, evenly spaced at intervals of 4 ft or less.

PIPE STACKING HEIGHTS

Coiled pipe is best stored as received in silo packs. Individual coils may be removed from the top of the silo pack without disturbing the stability of the remaining coils in the silo package.

Pipe received in bulk packs or strip load packs should be stored in the same package. If the storage site is flat and level, bulk packs or strip load packs may be stacked evenly on each other to an overall height of about 6 ft. For less level terrain, stacking height should be limited to about 4 ft.

Before removing individual pipe lengths from bulk packs or strip load packs, the pack must be removed from the storage stack and placed on the ground.

Individual pipes may be stacked in rows. Pipes should be laid straight, not crossing over or entangled with each other. The base row must be blocked to prevent sideways movement or shifting (Figure 7-5 and Table 7-1). The interior of stored pipe should be kept free of debris and other foreign matter.

EXPOSURE TO ULTRAVIOLET LIGHT AND WEATHER

PE pipe products are protected against deterioration from exposure to ultraviolet (UV) light and weathering effects. Color and black products are compounded with antioxidants, thermal stabilizers, and UV stabilizers. Color products use sacrificial UV stabilizers that absorb UV energy and are eventually depleted. In general, non-black products should not remain in unprotected outdoor storage for more than two years; however, some manufacturers may allow longer unprotected outside storage. After two years, the pipe should be recertified by additional testing or rejected. Black products contain at least 2 percent carbon black to protect the material from UV deterioration and are generally suitable for unlimited outdoor storage and for service on the surface or abovegrade.

COLD WEATHER HANDLING

Temperatures near or below freezing will affect PE pipe by reducing flexibility and increasing vulnerability to impact damage. Care should be taken not to drop pipe or fabricated fittings and to keep handling equipment and other things from hitting pipe. Ice, snow, and rain are not harmful to the material but may make storage areas more troublesome for handling equipment and personnel. Unsure footing and traction require greater care and caution to prevent damage or injury.

Walking on pipe can be dangerous. Inclement weather can make pipe surfaces especially slippery.

NOTE: Keep safety first on the jobsite; do not walk on pipe.

FIELD HANDLING

PE pipe is tough, lightweight, and flexible. Installation does not usually require high capacity lifting equipment. See Handling Equipment for information on handling and lifting equipment.

NOTE: To prevent injury to persons or property, safe handling and construction practices must be observed at all times. The installer must observe all applicable safety codes.

Pipe up to about 8-in. (219-mm) diameter and weighing roughly 6 lb per ft (20 kg per m) or less can usually be handled or placed in the trench manually. Heavier, larger diameter pipe will require appropriate handling equipment to lift, move, and lower the pipe. Pipe must not be dumped, dropped, pushed, or rolled into a trench.

NOTE: Appropriate safety precautions must be observed whenever persons are in or near a trench.

Coiled lengths and long strings of heat-fused PE pipe may be cold bent in the field. Field bending usually involves sweeping or pulling the pipe string into the desired

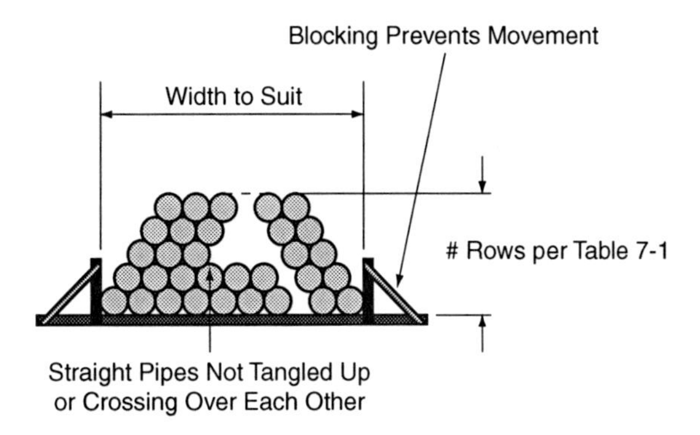

Figure 7-5 Loose pipe storage

Table 7-1 Suggested jobsite loose storage stacking heights for conventionally extruded pipe lengths

	Suggested Stacking Height*, Number of Rows	
Pipe Size	DR Above 17	DR 17 and Below
4	15	12
5	6	12
10	6	10
8	8	6
10	6	5
12	5	4
14	5	4
16	4	3
18	4	3
20	3	3
22	3	2
24	3	2
26	3	2
28	2	2
30	2	2
32	2	2
36	2	1
42	1	1
48	1	1
54	1	1
63	1	1

* Stacking heights based on 6 ft for level terrain and 4 ft for less level terrain.

bend radius, then installing permanent restraint, such as embedment around a buried pipe, to maintain the bend.

NOTE: Considerable force may be required to field bend the pipe, and the pipe may spring back forcibly if holding devices slip or are inadvertently released while bending. Observe appropriate safety precautions during field bending.

NOTE: Coils of PE pipe, especially sizes larger than 2 in., contain significant stored energy. When strapping bands are cut, coiled pipe will expand rapidly and forcibly with significant energy that can deliver hazardous impact, especially from the outer pipe end. Before cutting bands, coiled pipe 2 in. and larger should be installed in pay-out equipment that dispenses the coil at a controlled rate to help contain expansion. For all coil sizes, bands must be cut from the inside of the coil. Coils are typically banded in layers, and when cutting bands, the bands securing the outermost layers are cut first and one at a time. Successive interior layer bands are then cut one at a time. Persons cutting bands should use cutting tools designed for the purpose and should never place themselves or any appendages between the outside of the coil and the pay-out equipment. Significant pinch-point injury could result.

Chapter 8

Installation

GENERAL CONSIDERATIONS

PE pipe is tough, lightweight, and flexible. Installation does not usually require high capacity lifting equipment. See Chapter 7, *Transportation, Handling, and Storage of Pipe and Fittings,* for information on handling and lifting equipment.

To prevent injury to persons or property, safe handling and construction practices must be observed at all times. The installer must observe all applicable safety codes.

Heat fusion joining requires specialized equipment for socket-, saddle-, and butt-fusion and for electrofusion. Heat fusion joining may be performed in any season. During inclement weather, a temporary shelter should be set-up over the joining operation to shield heat fusion operations from precipitation and wind. Persons making heat fusion joints must be knowledgeable of procedural requirements for fusing in extremely cold weather. Most heat fusion equipment is not explosionproof. The equipment manufacturer's safety instructions must be observed at all times and especially when heat fusion is to be performed in a potentially volatile atmosphere. When installing PE pipe in a butt fusion machine, the pipe should not be bent against an open fusion machine collet or clamp. The pipe may suddenly slip out of the open clamp and cause injury or damage.

UNDERGROUND INSTALLATIONS

Buried installations of PE water distribution and transmission piping generally involve trench excavation, placing pipe in the trench, placing embedment backfill around the pipe, and then placing backfill to the required finished grade. Pipe application and service requirements, size, type, soil conditions, backfill soil quality, burial depth, and joining requirements will all affect the installation.

The care taken by the installer during installation will dramatically affect system performance. A high quality installation in accordance with recommendations and engineered plans and specifications can ensure performance as designed, while a low quality installation can cause substandard performance.

Additional information on pipe burial may be found in ASTM D2774[1] and from manufacturers.

Contaminated Soil

The selection of embedment materials and installation location is critical for water service and distribution piping in locations where the piping may be exposed to contamination such as low molecular weight hydrocarbons (petroleum fuels) or organic solvents or their vapors. PE pipe and elastomeric gaskets used in joining devices for other piping materials and PE pipes are permeable to these contaminants, and if permeation occurs, irreversible water service contamination may result.

It is strongly recommended not to install PE service, transmission, or distribution piping; permeable piping products; or permeable elastomeric gasket products in soils where contamination exists.

If installation through contaminated soils cannot be avoided, PE piping should be installed in a continuous, sealed casing that spans the contaminated area and extends into clean soil on both ends. The casing may be any piping material, including PE. The ends of the casing should be sealed to prevent backfill soil migration, and PE pipe entering and exiting a rigid (non-PE) casing must be protected against shear and bending loads. Another alternative is to remove contaminated soil.

Pipe Embedment Terminology

Embedment backfill materials that surround a buried pipe are defined by their function or location. (See Figure 8-1.)

Foundation. A foundation is required only when the native trench bottom does not provide a firm working platform or the necessary uniform and stable support for the installed pipe. If a foundation is installed, bedding is required above the foundation.

Initial backfill. This is the critical zone of embedment surrounding the pipe from the foundation to at least 6 in. over the pipe. The pipe's ability to support loads and resist deflection is determined by the quality of the embedment material and the quality of its placement. Within the initial backfill zone are bedding, haunching, primary, and secondary zones.

Bedding. In addition to bringing the trench bottom to grade if required, bedding levels out irregularities and ensures uniform support along the pipe length. Bedding is required when a foundation is installed, but a foundation may not be required to install bedding.

Haunching. The embedment under the pipe haunches (below the springline) supports the pipe and distributes the load. This area should be void-free to the extent possible to ensure that the compete side fill (haunching and primary initial backfill) provide proper support for the pipe. (The quality of the haunching backfill and primary initial backfill and their placement are the most important factors in limiting flexible pipe deformation.)

Primary initial backfill. This embedment zone provides primary support against lateral pipe deformation. It extends from the pipe invert to at least three-fourths of the pipe diameter height. Primary initial backfill extends to at least 6 in. above the pipe crown where the pipe will be continuously below normal groundwater levels.

Secondary initial backfill. Embedment material in this zone distributes overhead loads and isolates the pipe from any adverse effects from placing final backfill material. Where groundwater may rise over the pipe, the secondary initial backfill should be a continuation of the primary initial backfill.

Final backfill. Final backfill is not an embedment material; however, it should be free of large rocks, frozen clods, lumps, construction debris, stones, stumps, and any other material with a dimension greater than 8 in.

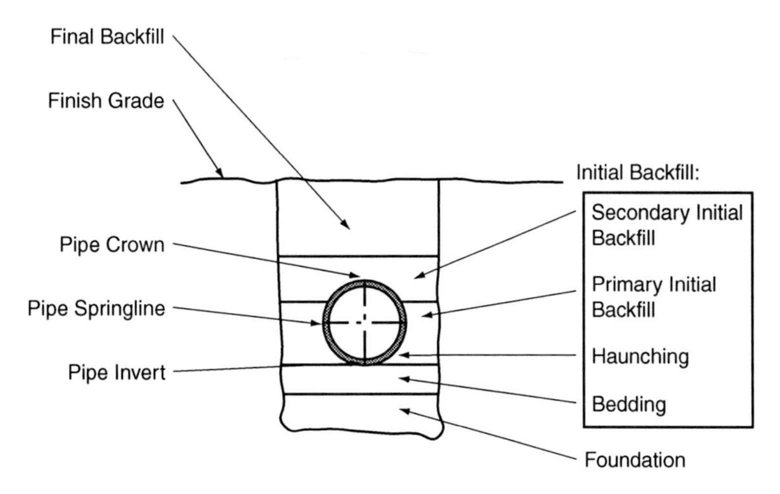

Figure 8-1 Trench construction and terminology

Engineered and Simplified Embedment

An engineered embedment is an underground installation where the pipe and embedment are designed to withstand earth loads and any static and dynamic loads that may be applied from the surface. (See Chapter 5, *External Load Design*.) Installations beneath or within easements for paved roads, highways, and railroads; under buildings or parking lots; and needing deep burials generally require an engineered embedment.

Plans, embedment soil specifications, and installation specifications prepared by a professional engineer must be observed when an engineered embedment is required. Select embedment materials may be required, but native soils that are sufficiently granular and free from cobbles, large rocks, clumps, frozen soil, and refuse and debris may frequently be acceptable. For an engineered embedment, the suitability of native soils for pipe zone embedment should be determined by a qualified engineer.

A simplified embedment may be suitable when the native underground environment and native soil meet the following guidelines. For example, simplified embedment is frequently suitable for rural transmission and distribution lines. Small diameter pressure pipes usually have adequate stiffness and are usually installed at such shallow depths that an engineered embedment is unnecessary. In some cases, a combination of engineered and simplified embedment may be applicable; that is, an engineered embedment would be installed only where loads are present, and a simplified embedment would be used for all other areas. In determining if an engineered embedment is required, the recommendations of a qualified engineer should be observed.

Guidelines for simplified embedment. A quality job can be achieved for most simplified embedment installations by observing the steps below. These guidelines apply where all of the following conditions are met:

- Pipe diameter of 24 in. or less
- DR equal to or less than 26
- Depth of cover between 2.5 ft and 16 ft
- Relatively stable groundwater elevation never higher than 2 ft below the surface
- Route of the pipeline is through stable soil

Stable soil is an arbitrary definition referring to soil that can be cut vertically or nearly vertically without significant sloughing or soil that is granular but dry (or dewatered) that can stand vertical to at least the height of the pipe. These soils must also possess good bearing strength. (Quantitatively, good bearing capacity is defined as a minimum unconfined compressive strength of 1,000 lb/ft^2 for cohesive soils or a minimum standard penetration resistance of 10 blows per ft for coarse grained soils.) Examples of soils that normally do not possess adequate stability for this method are mucky, organic, or loose and wet soils. Where the previous conditions are met, installation specifications in accordance with the following steps may be considered and in observance of applicable safety regulations.

Simplified embedment—step-by-step installation guidelines. The step-by-step installation guidelines are general guidelines for simplified embedment installation of ANSI/AWWA C906[2] piping. Other satisfactory methods or specifications may be available. This information should not be substituted for the judgment of a professional engineer in achieving specific requirements.

Trenching. Trench collapses can occur in any soil and account for a large number of worker deaths each year. In unbraced or unsupported excavations, proper attention should be paid to sloping the trench wall to a safe angle. All applicable codes should be consulted. All trench shoring and bracing must be kept above the pipe. (See the Trenching section.) The length of open trench required for fused pipe sections should be such that bending and lowering the pipe into the ditch does not exceed the manufacturer's minimum recommended bend radius and result in kinking. (See the Cold (Field) Bending section.) The trench width at pipe grade should be equal to the pipe outer diameter (OD) plus 12 in.

Dewatering. For safe and proper construction, the groundwater level in the trench should be kept below the pipe invert. This can be accomplished by deep wells, well points, or sump pumps placed in the trench.

Bedding. Where the trench bottom soil can be cut and graded without difficulty, pressure pipe may be installed directly on the prepared trench bottom. For pressure pipe, the trench bottom may undulate but must support the pipe smoothly and be free of ridges, hollows, and lumps. When encountering rocks, boulders, or large stones that may cause point loading on the pipe, remove the rocks, boulders, and large stones and pad the trench bottom with 4 to 6 in. of tamped bedding material. Bedding should consist of free flowing material such as gravel, sand, silty sand, or clayey sand that is free of stones or hard particles larger than 1/2 inch.

Pipe embedment. Figure 8-1 shows trench construction and terminology. Haunching and initial backfill are considered trench embedment materials. The embedment material should be a coarse-grained soil, such as gravel or sand, or a coarse-grained soil containing fines, such as a silty sand or clayey sand. The embedment material particle size should not exceed 1/2 in. for 2- to 4-in. pipe, 3/4 in. to 1 in. for 6- to 8-in. pipe, and 1 1/2 in. for all other sizes. At the discretion of a qualified engineer, embedment materials for 8-in. and smaller pipes may contain larger particles up to 1 1/2 in. where the larger particles are 5 percent or less of the mix. Where the embedment is angular, crushed stone may be placed around the pipe by dumping and slicing with a shovel. Where the embedment is naturally occurring gravels, sands, and mixtures with fines, the embedment should be placed in lifts, not exceeding 6 in. in thickness and then tamped. Tamping should be accomplished with a mechanical tamper. Compact to at least 85 percent Standard Proctor Density as defined in ASTM D698[3]. Under streets and roads, compaction should be increased to 95 percent Standard Proctor Density.

Leak testing. If a leak test is required, it should be conducted in accordance with ASTM F2164[4] after the embedment material is placed.

Trench backfill. The final backfill may consist of the excavated material, provided it is free from unsuitable matter such as large lumps of clay, organic material, boulders, or stones larger than 8 in., or construction debris. Where the pipe is located beneath a road, place the final backfill in lifts as previously mentioned and compact to 95 percent Standard Proctor Density.

Whether the installed embedment is engineered or simplified, long-term piping service and performance are maximized by proper placement and compaction of embedment materials. Pipe zone embedment provides encapsulating soil support for flexible PE pipe. Poorly placed embedment that allows large unsupported areas and voids around the pipe can result in excessive deformation and unnecessary stresses, especially with higher DR (thinner wall) pipes. Embedment that is simply dumped into the trench but not worked under the pipe haunches and compacted may not provide adequate support and may lead to reduced performance or premature failure. Quality embedment installation that provides uniform soil support all around the pipe ensures long-term performance.

Trenching

In stable ground, minimum trench width, B_d, will vary by the pipe diameter as illustrated in Figure 8-2 and Table 8-1. The trench must be wide enough to place and compact backfill soils in the haunch areas below the pipe springline.

To minimize the load on the pipe, the maximum trench width should not exceed the minimum trench width by more than 18 in. plus the thickness of any sheeting, shoring, or shielding, unless approved by the engineer. For trenches containing multiple pipes, the distance between parallel pipes should be the same as the clearance distance between the pipe and the trench wall. (See Table 8-1.)

Depending on trench soil stability and depth, trench sides above the pipe crown may need to be sloped or stepped as illustrated in Figure 8-2. When trenching in unstable ground, the trench width above the pipe crown should be sloped and/or widened. Trench sidewall bracing such as trench shielding, sheeting, or a trench box should always be used wherever required by applicable safety regulations.

When using a trench box, a trench offset should be excavated at a depth between the pipe crown and one quarter of the pipe diameter below the pipe crown; then the trench box should be installed on the offset shelf. Further excavation of the pipe zone trench down to the foundation grade should be performed within the protection of the trench box. (See Figure 8-3.)

Accurate leveling of the trench bottom is usually not necessary for pressure piping systems such as raw or potable water mains or sewage force mains. The trench bottom must support the pipe continuously and be free from ridges, hollows, lumps, and the like. Any significant irregularities must be leveled off and/or filled with compacted embedment backfill. Leveling may also be a consideration if the pipeline must be drained and to help avoid high spots where air binding could occur. When encountering rocks, boulders, or large stones that may cause point loading on the pipe, remove the rocks, boulders, and large stones and pad the trench bottom with 4 in. to 6 in. of tamped bedding material. If the trench bottom is reasonably uniform, and the soil is stable and free of rock, foundation or bedding may not be required.

The pipe should be laid on a stable foundation. Where water is present in the trench or where the trench bottom is unstable, excess water should be removed before laying the pipe. Groundwater should be lowered to below the level of the bedding material. During dewatering, care should be taken not to remove sand or silt and not to displace foundation or bedding soil material.

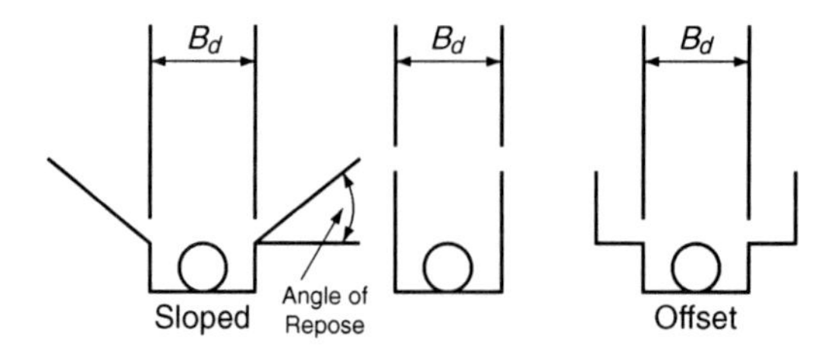

Figure 8-2 Trench width

Table 8-1 Minimum trench width

Nominal Pipe OD, *in.*	Minimum Trench Width, B_d, *in.*	Parallel Pipe Clearance, *in.*
<3	12	4
3–24	Pipe OD + 12	6
>24–63	Pipe OD + 24	12

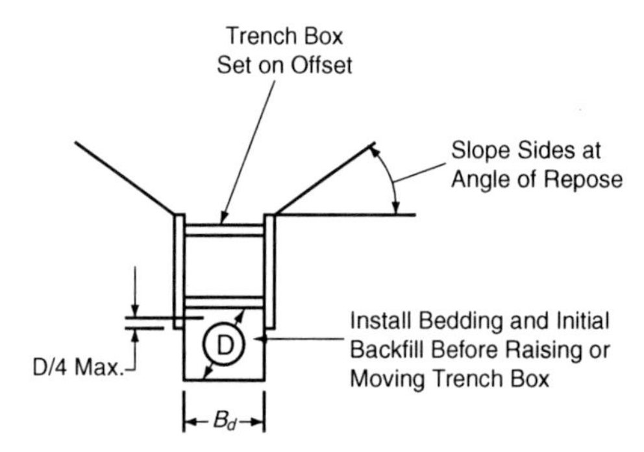

Figure 8-3 Trench box installation

Where an unstable trench bottom exists such as in mucky or sandy soils with poor bearing strength, trench bottom stabilization may be required. Stabilization may be provided by excavating the trench below the pipe bottom grade and installing a foundation and bedding, or bedding alone, to the pipe bottom grade. When required, the minimum foundation thickness is 6 in. When bedding and foundation are both required, the minimum bedding thickness is 4 in. Without a foundation, the minimum bedding thickness is 6 in.

For an engineered embedment, all materials used for bedding, haunching, and primary and secondary backfill should be installed to at least 85 percent Standard Proctor Density. Mechanical compaction, which may be as simple as shovel slicing Class I material, is usually required to achieve 85 percent Standard Proctor Density. Native soils may require additional work to achieve necessary compaction.

When the pipe is laid in a rock cut or stony soil, the trench should be excavated at least 6 in. below pipe bottom grade and brought back to grade with compacted bedding. Ledge rock, boulders, and large stones should be removed to avoid point contacts and to provide a uniform bed for the pipe.

The approximate length of open trench required to lay long strings of ANSI/AWWA C906 PE pressure pipe may be determined from Eq 8-1:

$$L_{OT} = 1.75\sqrt{100 + 6H_{OT}(D_o - H_{OT})} \quad \text{(Eq 8-1)}$$

Where:

L_{OT} = trench length, ft

H_{OT} = trench depth, ft

D_o = pipe outside diameter, in.

Eq 8-1 is applicable when cover above the pipe crown equals or exceeds one pipe diameter.

Placing Pipe in the Trench

ANSI/AWWA C906 pipe up to about 8-in. diameter and weighing roughly 6 lb per ft or less can usually be placed in the trench manually. Heavier, larger diameter pipe will require appropriate handling equipment to lift, move, and lower the pipe into the trench. Pipe must not be dumped, dropped, pushed, or rolled into the trench. Appropriate safety precautions must be observed whenever persons are in or near the trench. Recommendations for handling and lifting equipment are discussed in Chapter 7, *Transportation, Handling, and Storage of Pipe and Fittings.*

Black PE pipe subjected to direct sunlight or warm ambient air temperature may become warmer than the ground temperature. When placed in a trench, the pipe will contract in length as it cools to the soil temperature. If the pipe is connected to subsurface structures before it has cooled sufficiently, excessive pull-out forces could develop. Allow the pipe to cool prior to making a connection to an anchored joint, flange, or a fitting that requires protection against excessive pull-out forces. Soil embedment can be placed around the pipe and up to 6 in. over the top of the pipe to facilitate cooling.

Cold (Field) Bending

PE pipe may be cold bent in the field without affecting the working pressure rating. The minimum permissible long-term bend radius is determined by pipe diameter and dimension ratio unless a fitting or a flange connection is either present or will be installed in the bend. (See Table 8-2 and Figure 8-4.)

Field bending usually involves excavating the trench to the desired bend radius, sweeping or pulling the pipe string into the required bend, and placing it in the trench. Temporary restraints may be required to bend the pipe and to maintain the bend while placing the pipe in the trench and placing initial backfill. Temporary blocks or restraints must be removed before installing final backfill, and any voids must be filled with compacted initial backfill material. Equipment such as spreader bars and slings can be helpful in distributing the bending load over a larger area of the pipe during field bending. Distributing the bending load helps avoid kinking.

PE will typically bend without great effort to about a 70-pipe diameter or greater bend radius. Considerable force may be required to produce tighter field bends, and the pipe may spring back forcibly if the restraints slip or are inadvertently released while bending. Appropriate safety precautions should be used during field bending. If appropriate equipment and restraints are not available for field bending, directional fittings should be installed.

Table 8-2 Minimum cold (field) bending radius (long-term)

Pipe DR	Minimum Cold Bending Radius†
≤9	20 times pipe OD
>9–13.5	25 times pipe OD
>13.5–21	27 times pipe OD
>21	30 times pipe OD
Fitting or flange present or to be installed in bend*	100 times pipe OD

* Observe the minimum cold bending radius for a distance of about 5 times the pipe diameter on either side of the fitting location.

† Values include a safety factor against kinking of at least 2.

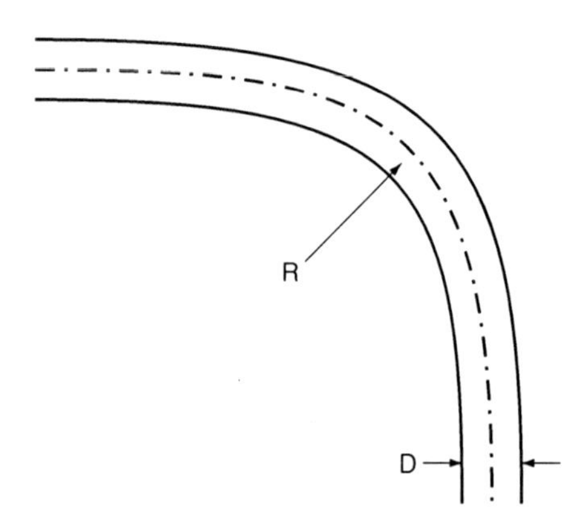

Figure 8-4 Bend radius

Installing Fabricated Fittings

To avoid field damage, large diameter (16-in. IPS and above) fabricated directional fittings, such as elbows, tees, wyes, and crosses, should not be joined to more than one pipe before placement in the trench. The remaining outlet connections are made after placement in the trench with flanges, mechanical couplings, or electrofusion couplings. Butt fusion can be performed in the trench, but placing and removing the butt fusion machine in the trench must be done such that the piping is not disturbed. Connecting pipes to more than one fitting outlet, then attempting to lift, move, and then lower the assembly into the trench frequently causes fitting breakage and is not recommended.

Pipe Embedment Soils for Engineered Embedment

Preferred engineered embedment soils for ANSI/AWWA C906 PE pipes are angular Class I and Class II gravels and sands classified as meeting soil types GW, GP, SW, or SP and dual classifications beginning with one of these symbols as defined in ASTM D2487[5]. These materials should be used for bedding, haunching, and for primary and secondary initial backfill. The maximum particle size should be limited to 1/2 in. for pipes to 4-in. diameter, 3/4 in. for pipes 6-in. to 8-in. diameter, 1 in. for pipes 10-in. to 16-in. diameter, and 1 1/2 in. for larger pipes. (See ASTM D2774.)

Class III materials may be used in the engineered embedment zone only when specified by the engineer, and if allowed, they must be compacted to at least 90 percent

Table 8-3 Embedment soil classification

Embedment Backfill Class	Soil Description – Pipe Embedment Material*
Class I	Manufactured angular, granular material with little or no fines. Angular crushed stone, particle size ¼ in. to 1½ in., including materials of regional significance such as marl, coral, crushed shells, cinders, slag, etc.
Class II	Coarse-grained soils with little or no fines—GW, GP, SW, SP† containing less than 12% fines
Class III	Coarse-grained soils with fines—GW, GP, SW, SP containing more than 12% fines
Class IVa	Fine-grained soils (LL‡ < 50); soils with medium to no plasticity—CL, ML, ML-CL with more than 25% coarse-grained particles
Class IVb	Fine-grained soils (LL > 50); soils with medium to high plasticity—CH, MH, CH-MH. Fine-grained soils (LL < 50); soils with medium to no plasticity—CL, ML, ML-CL with less than 25% coarse-grained particles

* ASTM D2487, USBR Designation E-3.

† Or any borderline soil beginning with one of these symbols (e.g., GM-GC, GC-SC, etc.).

‡ LL = liquid limit.

Standard Proctor Density. Class IVa and Class IVb materials are not preferred and should be used only with the specific approval of a geotechnical soils engineer. (See Chapter 5, *External Load Design*.)

Embedment soils are defined and classified in accordance with ASTM D2487, ASTM D2488[6], and USBR Designation E-3. Table 8-3 provides information on embedment materials suitable for ANSI/AWWA C906 PE pipe.

For a given density or compaction level, Class III and Class IVa soils provide less support than Class I or Class II soils. Placement of Class III and Class IVa materials is difficult, soil moisture content must be carefully controlled, and significantly greater effort is required to achieve the necessary compacted density for pipe support.

Embedment Backfilling

Where an engineered backfill is required, underground ANSI/AWWA C906 pipes should be installed in accordance with ASTM D2774.

Whether or not an engineered embedment is required, voids in the haunch areas are undesirable. The haunch areas should be completely filled and void free to the extent possible. For the lower half of the haunch area, materials should be shoveled evenly into the area on both sides of the pipe in layers not more than 4 in. thick and compacted. Layers can then be increased to 6 in., and flat-tamping tools can be used. (See Figure 8-5.)

After haunching, primary and secondary initial backfill materials should be placed in 6-in. layers and compacted with flat tamping tools. If mechanical tampers are used, care should be taken not to damage the pipe. If trench wall side bracing has been used, the bracing should be lifted progressively for each layer.

Consult with the pipe manufacturer or a qualified engineer on other embedment methods such as saturation and vibration or flowable fills.

Thrust Blocks

ANSI/AWWA C906 PE pressure piping systems must be assembled with fully restrained joints or with partially restrained joints AND external joint restraints. **ANSI/AWWA C906 pressure piping systems that are joined by heat fusion, electrofusion,**

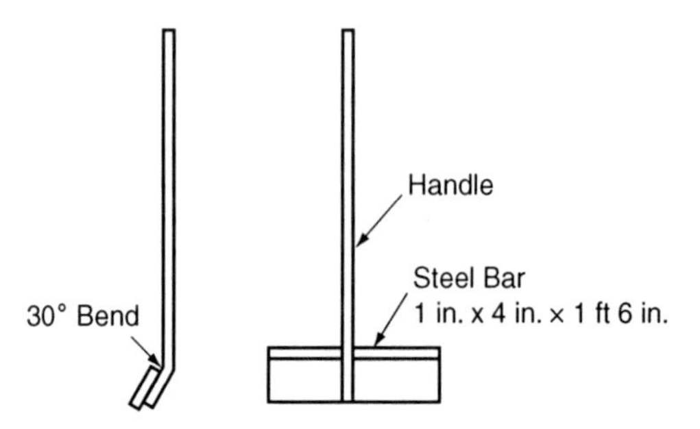

Figure 8-5 Example haunch tamping tool

flanges, and MJ adapters are fully restrained and do not require external joint restraints or thrust block joint anchors.

In-Line Anchoring (Poisson Effects)

When ANSI/AWWA C906 piping is connected in-line to unrestrained joint piping or components such as bell and spigot joint PVC or ductile iron, unrestrained joints in the transition area should be protected against pullout disjoining with external mechanical restraint, or in-line anchoring, or a combination of restraint and in-line anchoring. This protection is different from thrust blocks at directional fittings, which are not effective against Poisson effect pullout. Conventional thrust blocks are intended to contain fluid flow thrust forces that would push the fitting off the end of the pipe, but thrust blocks cannot counteract forces that would pull the pipe end out of the joint. As well, snaking pipe in the trench is generally not effective.

When pipes made from ductile materials are pressurized, the diameter expands slightly and the length decreases in accordance with the Poisson ratio of the material. With unrestrained bell and spigot joined lengths, the effect is limited to the individual pipe lengths, but with fully restrained piping such as fusion-joined ANSI/AWWA C906 pipe, the effect is cumulative over the entire restrained length of pipe. When ANSI/AWWA C906 piping is connected to unrestrained mechanical couplings or bell and spigot joint PVC or ductile iron, Poisson effect pipe shortening can cause pullout disjoining of unrestrained in-line joints in the transition area. To prevent Poisson effect pullout disjoining, protection should be provided by installing external joint restraints at unrestrained bell and spigot joints, or by installing an in-line anchor in the PE pipeline, or by a combination of both techniques.

Pullout Prevention Techniques

The transition region where a long string of ANSI/AWWA C906 pipe is connected in-line to unrestrained piping can extend several joints into the non-PE pipe system. A restrained connection at the transition joint can transmit Poisson shortening to the next-in-line unrestrained joint in the non-PE pipe. Typical pullout prevention techniques include restraining the transition connection and installing an in-line anchor in the PE pipe close to the transition connection or restraining several non-PE pipe joints down line from the transition connection. Figures 8-6 and 8-7 illustrate typical pullout prevention techniques. Examples of in-line anchors are illustrated in Figures 8-8 and 8-9.

Determining Poisson Effect Pullout Force

Poisson effect pipe shortening will occur whenever the pipe is pressurized. Because internal pipe pressures are higher during pressure testing and surge events, Poisson

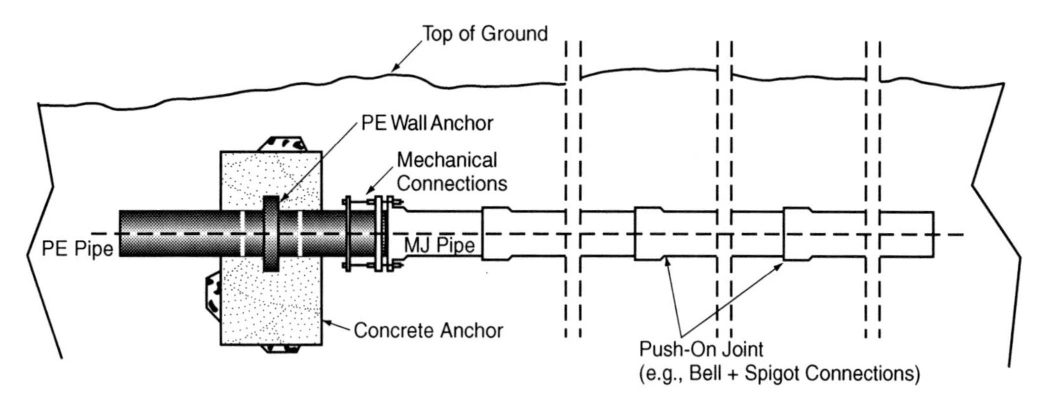

Figure 8-6 Pullout prevention technique

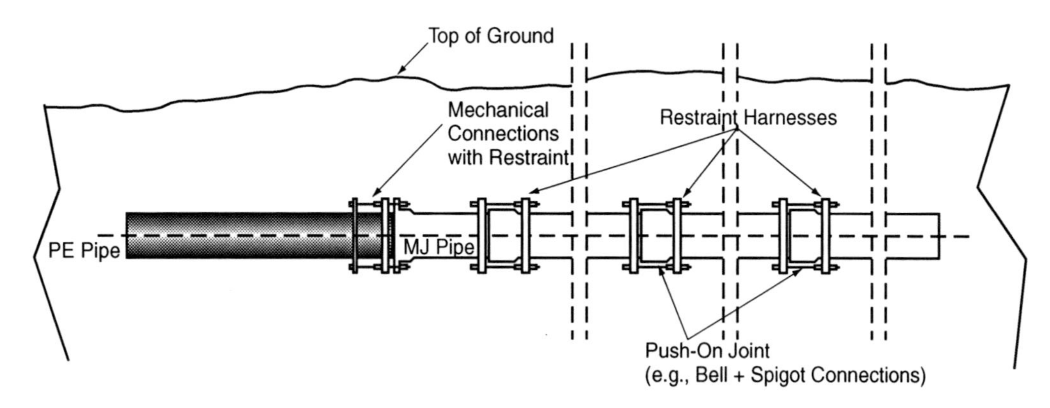

Figure 8-7 Pullout prevention technique

effect pipe shortening can be greater at these times compared to normal steady pressure operation. A qualified engineer should determine the Poisson effect pullout force conditions that are appropriate for the application, and then determine the appropriate techniques to protect unrestrained in-line mechanical connections against disjoining from Poisson effect pullout forces.

For a given PE pipe diameter and DR, approximate Poisson effect pullout force may be determined by multiplying the end area of the ANSI/AWWA C906 pipe by the product of the internal pressure hoop stress and the appropriate Poisson ratio. (See Eq 8-2.)

$$F = S_P \mu \pi D_M^2 \left[\frac{1}{DR} - \frac{1}{DR^2}\right] \qquad \text{(Eq 8-2)}$$

Where:

F = pullout force, lb

S_P = internal pressure hoop stress, lb/in.^{2}

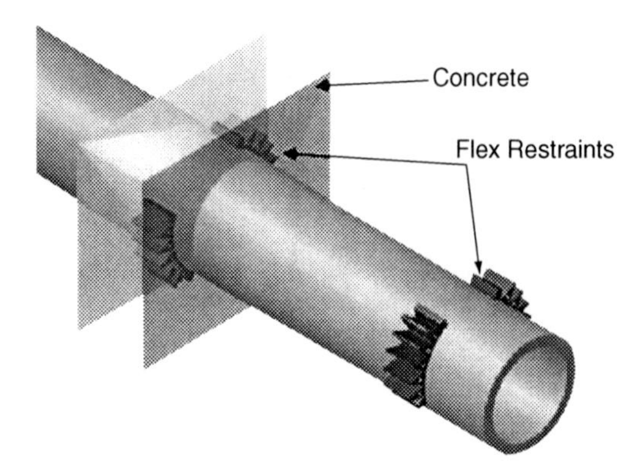

Figure 8-8 In-line anchor using flex restraint

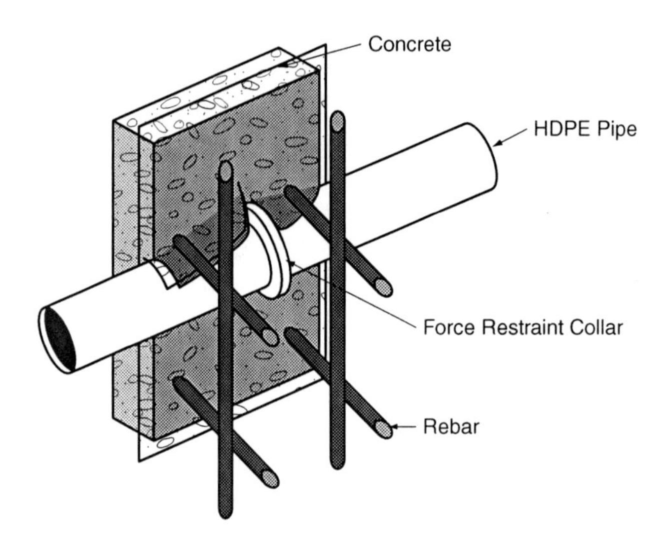

Figure 8-9 In-line anchor using integral pipe collar (wall anchor)

$$S_P = \frac{P(DR-1)}{2} \qquad \text{(Eq 8-3)}$$

Where:

P = internal pressure, lb/in.^{2}

DR = dimension ratio

μ = Poisson ratio (for PE, 0.45 for long-term stress; 0.35 for short-term stress)

π = Pi (approximately 3.142)

D_M = pipe mean diameter, in.

The value of P will vary depending on the internal pressure conditions. When operating at steady maximum rated internal pipe pressure, $P = PC$. During leak testing at maximum test pressure, $P = 1.5 \times PC$, and during an emergency surge event for pipe operating at PC or $P = P_{(MAX)(OS)}$.

Table 8-4 presents approximate Poisson effect pullout forces for selected sizes of ANSI/AWWA C906 pipe while operating at rated system internal pressure, during

Table 8-4 Approximate Poisson effect pullout force

ANSI/AWWA C906 Pipe Size (DR 11) *(in.)*	Approximate Pullout Force, *F, lb* (a)		
	Operating at Full Rated Pressure $P = PC$ (b)	During Pressure Tests at 150% of Rated Pressure $P = 1.5 \times PC$ (c)	Operating at Full Rated Pressure Plus Maximum Allowable Occasional Surge Pressure $P = P_{(MAX)(OS)}$ (d)
4	1,892	2,208	3,364
6	4,102	4,786	7,293
8	6,953	8,112	12,361
10	10,801	12,602	19,202
12	15,195	17,727	27,013
16	23,928	27,916	42,539

(a) Values for water at 80°F (27°C) and below.

(b) Rated pressure for DR 11, Class 160 = 160 psi. Pullback force determined using long-term Poisson ratio of 0.45.

(c) Pullback force determined using short-term Poisson ratio of 0.35.

(d) Total pressure in pipe during surge event = 160 psi steady pressure + 160 psi surge pressure = 320 psi. Values determined by combining pullback force for steady pressure (long-term Poisson ratio of 0.45) plus pullout force for surge event (short-term Poisson ratio of 0.35).

leak testing at 150 percent of rated system pressure, and during a severe water hammer event while operating at steady pressure that causes a pressure surge to 200 percent of rated system pressure.

Other longitudinal forces from thermal expansion and contraction, fluid thrust, or installation are not incorporated into table values.

Controlling Shear and Bending Loads at Transition Areas

ANSI/AWWA C901[7] and C906 pipes that enter or exit a casing or pass through a structure wall, such as a building wall, vault, or access hole, or land-marine transition areas, must be protected against shear and bending loads that can develop from settlement and embedment consolidation waves, currents, or tidal flows.

A compacted foundation and compacted bedding should be installed below the pipe where it exits the casing or structure as illustrated in Figure 8-10. At a casing entry or exit, the pipe should be wrapped with an elastomeric sheet material; then the annulus between the pipe and the casing should be sealed either mechanically or with a cement grout. The seal prevents backfill soil migration into the annulus.

Where PE pressure pipe is flanged at a wall such as a building or vault wall, a structural support as illustrated in Figure 8-11 is recommended to prevent shear and bending loads. Within the clamp, the pipe is protected against chafing by wrapping it with an elastomeric sheet.

Where ANSI/AWWA C906 pipe or fittings are joined to valves, hydrants, other heavy devices, or rigid pipes, a support pad as illustrated in Figure 8-12 should be provided below the device or rigid pipe and for at least two pipe diameters length under the connecting pipes. Support pad materials should be at least compacted Class I or II soil; cement stabilized Class I, II, or III soils; or poured concrete. Embedment soils around the connecting pipes, the device, and in any bell holes must be compacted.

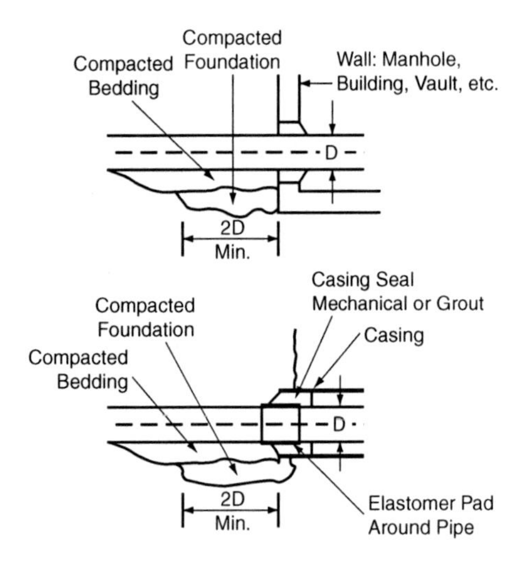

Figure 8-10 Controlling shear and bending

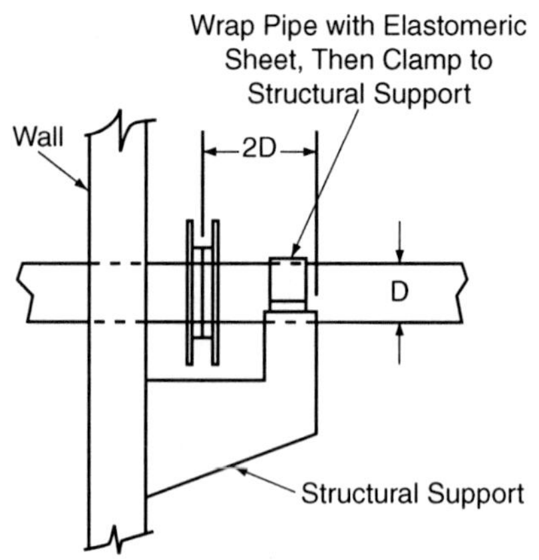

Figure 8-11 Flange support at wall

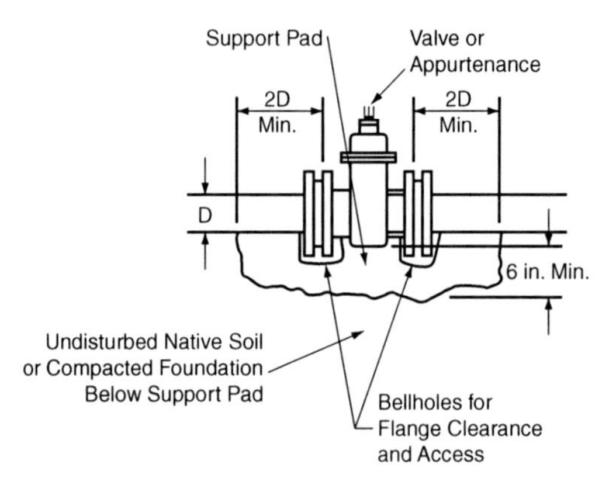

Figure 8-12 Appurtenance support pad

Locating Tapes or Wires

PE pipes are nonmetallic, so once buried, metal detector type locators are ineffective. To facilitate locating buried pipe, metallic locating tapes or 12-14 AWG copper wires can be placed in the trench. Locating tapes or wires are usually placed slightly above the pipe between the initial and final backfill. (If a locating wire is placed on the pipe, a lightning strike may cause the locating wire to become hot enough to damage the pipe.)

Final Backfilling

In general, final backfill may be material excavated from the trench provided it is free of unsuitable matter such as lumps, stones, frozen clods, construction debris, boulders, and other materials exceeding 8 in. in their longest dimension.

Where the trench is subject to surcharge loads such as H20 or E80 live loads, or building foundations or footings, or paved parking or storage areas, final backfill should be an angular Class I or Class II granular material, compacted to at least 95 percent Standard Proctor Density or as specified by the engineer.

Backfill Stabilizing Agents

In some regions, soil may be stabilized by the addition of a pebble, lime, and starch-base polymer mixture. This mixture reacts with soil moisture and can generate temperatures in excess of 200°F (93°C). PE pipe coming into direct contact with the reacting mixture can be weakened or damaged.

Therefore, unreacted stabilizer and unreacted stabilized soil must not come into contact with PE pipe. Mixing soil and stabilizing agents as the soil is excavated from the trench will usually allow sufficient time for reaction and cooling. After mixing, the stabilized soil must be allowed to react and cool completely before placing it around PE pipe.

Burrowing Animal Control

PE contains nothing to attract burrowing rodents or insects; however, burrowing animals may find it easy to dig in trench backfill soils and may occasionally damage 4-in. and smaller PE pipes in their path. Larger pipes are too great an obstruction and are generally free from such attacks.

The possibility of burrowing animal damage may be reduced by a) using commercially available repellent sprays in the ditch; b) installing smaller lines below the animal's normal activity area, typically deeper than 3 ft; c) using larger diameter pipes; d) installing embedment zone backfill that is difficult for the animals to penetrate such as angular crushed stone or flowable fill; or e) installing the pipe or tubing within a shielding tube.

SPECIAL INSTALLATION TECHNIQUES

Because of its flexibility and the high integrity of properly made butt fusion joints, special installation techniques may be employed to install ANSI/AWWA C901 and C906 PE pipe. Special installation techniques include plowing, planting, pulling pipe into a narrow trench, pipe bursting, insertion renewal, horizontal boring, and directional boring. These techniques minimize excavation by making a tight-fitting trench cut or hole for the pipe, and either pulling or placing the pipe in the cut. They require suitable native soil conditions that are free of large rocks and except for directional boring, are generally limited to shallower depths.

Special installation techniques that involve pulling can apply potentially damaging short-term tensile stresses that must be controlled within safe levels for the pipe. All special installation techniques that involve pulling should be reviewed for allowable tensile load (safe pull strength), and protective devices such as weak links or breakaway devices may be needed. Pull-in installations should be visually monitored at both ends of the pull to ensure continuous movement at both ends. Visual monitoring is necessary because tensile yield and tensile break strengths are about the same. If the pipe yields, it will stretch to the breaking point with virtually no change in pulling force, that is, pull force gauges cannot identify that the pipe has yielded. The only indication will be that the trailing end stops while the pulling end continues to move.

Pulling force protective devices, pulling force monitoring, and visual movement monitoring are especially important with smaller piping where pulling equipment can exceed the pipe's allowable tensile load rating. Where pulling equipment cannot exceed the allowable tensile load rating, weak links or breakaway devices are not required.

Allowable Tensile Load (Safe Pull Strength)

When ANSI/AWWA C906 pipe is subjected to a significant short-term pulling stress, the pipe will stretch somewhat before yielding. However, if the pulling stress is limited to about 40 percent of the yield strength, the pipe will usually recover undamaged to its original length in a day or less after the stress is removed. Eq 8-4 gives an allowable tensile load (ATL) value that is 40 percent of the pipe nominal yield strength.

ATL (safe pull strength) may be determined by

$$ATL = \pi D_o^{\,2} f_Y \, f_T \, T_Y \left(\frac{1}{DR} - \frac{1}{DR^2} \right) \qquad \text{(Eq 8-4)}$$

Where:

ATL = allowable tensile load, lb

D_o = pipe outside diameter, in.

f_Y = tensile yield design (safety) factor (Table 8-5)

f_T = time under tension design (safety) factor (Table 8-5)

T_Y = pipe tensile yield strength, lb/in.^{2} (Table 8-6)

DR = pipe dimension ratio

Pipe yield strengths may be estimated by using the values from Table 3-8. Unlike more brittle materials, PE pipe materials stretch greatly, at least 400 percent, between tensile yield and tensile break. Tensile yield to break elongations of 800 percent for HDPE and 1,000 percent for MDPE are common. See ASTM F1804[8] for additional information on allowable tensile load.

Because pull-in loads will cause the pipe to stretch, the leading end should be pulled past the termination point by 4 to 4½ percent of the total pulled-in length, and the trailing end should be left long by the same amount. Final tie-ins should be made a day after the pull to allow the pipe to recover from the pulling stress and contract to its original prepull length. The extra length at both ends ensures that the pipe will not recede past the tie-in points as it recovers from the pull.

Plowing and Planting

Plowing and planting involve cutting a narrow trench and feeding the pipe into the trench through a shoe or chute fitted just behind the trench cutting equipment. Trench cuts for pipes around 1½-in. IPS and smaller are frequently made with vibratory plows. Larger sizes use wheel- or chain-type trenchers with semicircular cutters. The trench width should be only slightly larger than the pipe outside diameter.

The shoe or chute should feed the pipe into the bottom of the cut. The short-term pipe bending radius through the shoe may be tighter than the long-term cold bending radius in Table 8-2, but it must not be so tight that the pipe kinks. Table 8-7 presents minimum short-term bending radii for applications such as plowing and planting and submergence during marine installations. The pipe's path through the shoe or chute should be as friction free as practicable.

Pipe is usually fed over the trenching equipment and through the shoe or chute from coils or straight lengths that have been butt fused into a long string. Pipe up to 12-in. IPS has been installed using this method.

Table 8-5 Recommended design factors

Factor	Parameter	Recommended Value		
f_Y	Tensile yield design factor*	0.40		
f_T	Time under tension design factor	1.0 for up to 1 hr	0.95 for up to 12 hr	0.91 for up to 24 hr

* Design and safety factors are the inverse of each other. Multiplying by a 0.40 design factor is the same as dividing by a 2.5 safety factor.

Table 8-6 Approximate tensile yield strength values

	Approximate Tensile Yield Strength, T_Y, at Pipe Temperature			
Material	73°F (23°C)	100°F (38°C)	120°F (49°C)	140°F (60°C)
PE 2406	2,600 lb/in.^{2} (17.9 MPa)	2,365 lb/in.^{2} (16.3 MPa)	1,920 lb/in.^{2} (15.4 MPa)	1,640 lb/in.^{2} (11.0 MPa)
PE 3408	3,200 lb/in.^{2} (22.1 MPa)	2,910 lb/in.^{2} (17.4 MPa)	2,365 lb/in.^{2} (13.7 MPa)	2,015 lb/in.^{2} (14.3 MPa)

Table 8-7 Minimum short-term bending radius

Pipe Dimension Ratio	Minimum Short-Term Bending Radius
≤9	10 times pipe OD
>9–13.5	13 times pipe OD
>13.5–21	17 times pipe OD
>21	20 times pipe OD

Horizontal Boring

Horizontal boring or road boring is usually performed to install a pipeline below existing roadways or structures where opening a trench may be impractical or undesirable. Typically, entry and exit pit excavations are required. Tunneling directly across and under the structure makes the bore.

Road bores are usually performed using a rotating auger within a steel casing. The auger projects just ahead of the casing, and the auger and casing are advanced together across to the exit pit. If a casing is being installed, either the auger casing is left or a new casing is installed by pulling it in from the exit pit while withdrawing the bore casing.

ANSI/AWWA C906 pipe may be installed through a casing or directly in the borehole. For information on sealing the end of the casing, see the section Controlling Shear and Bending Loads at Transition Areas in this chapter.

When installed in a casing, pipe does not require centering spacers (centralizers) for electrical isolation to a metal casing. PE is nonconductive and will not affect casing cathodic protection. Grouting the casing annulus is not required.

Allowing the pipe to snake inside the casing can usually accommodate minor thermal length changes of the PE pipe in the casing. If used, centering spacers will force thermal expansion thrust loads to the pipe ends, which may weaken or break casing end seals.

When installing pipe either directly in the borehole or in a casing, joining to the installed pipe should be considered. Generally, the trailing end may be joined to the system by any appropriate method. However, the leading end may need to be a restrained mechanical joint, or electrofusion may need to be used. If a casing is large enough to allow a flange adapter to pass, a split backup ring as illustrated in Figure 8-13 may be used for the flange joint.

When installing smaller diameter pipes directly in a borehole, soil friction around the pipe may result in significant pulling forces. The allowable tensile load, Eq 8-4, for the pipe must not be exceeded.

Pipe Bursting

Pipe bursting employs a pulling head that fractures and enlarges an original (host) pipe and then pulls a new PE pipe of the same or larger size into the fractured host pipe. In some cases, a sleeve pipe is installed inside the fractured host pipe before installing the PE pipe. The sleeve pipe protects the PE pipe when fracturing the host pipe leaves sharp edges that could cut the surface of the PE pipe.

Insertion Renewal

In insertion renewal, which is similar to sliplining, a new, smaller PE pipe is pulled inside a host pipe. Insertion renewal can restore leak tightness and pressure capacity when excessive leakage or reduced pressure capacity has occurred because of age or corrosion effects.

Horizontal Directional Drilling

Horizontal directional drilling (HDD) uses directional drilling techniques to guide a drill string along a bore path around obstacles such as under rivers or lakes or through congested underground infrastructure. As with horizontal boring, horizontal directional drilling may be used to install a casing or to directly install long strings of ANSI/AWWA C901 or C906 pipe.

As the hole is bored, a steel drill string is extended behind a cutting head. Drilling mud is used to cool the cutter, flush excavated soil from the borehole, and lubricate the borehole. At the end of the bore path, the drill string is angled upward and through the surface. The cutting head is removed and a backreamer attached. The pipe string is attached to the backreamer through a weak-link or breakaway device (if required). As the drill string is withdrawn to the drilling rig, the backreamer enlarges the borehole and the pipe string is drawn in. As with any pipe pulling technique, the movement of the pipe string must be visually monitored, and the pulling load on the PE pipe (Eq 8-4) must not exceed the pipe's allowable tensile load rating. Additional information on horizontal directional drilling is available in ASTM F1962[9].

MARINE INSTALLATIONS

This section provides basic information about the design and installation of PE marine pipeline systems. It is intended to provide general guidance for typical installations. It does not provide complete information for all possible marine installations. Additional information on marine pipeline installations is available in the Marine Installations chapter of the *PPI Handbook of Polyethylene Piping*[10]. Because marine pipeline design and installation can be complex, the services of qualified professionals having experience with marine pipeline design and installation are strongly recommended.

PE pipe is frequently used for fresh water and saltwater marine applications where the pipeline runs along or across the water body, or for outfalls or intakes. PE pipe may be submerged into a trench in the riverbed, lakebed, or seabed and backfilled, submerged onto the bed and backfilled, exposed on the bed, submerged above the bed at some distance below the surface, or floated at or above the surface. Exposed

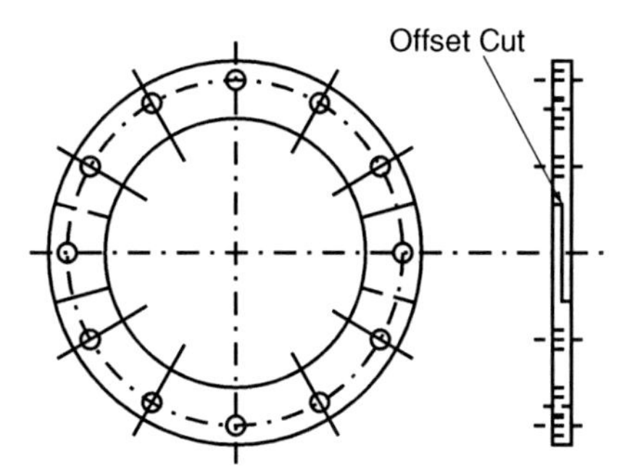

Figure 8-13 Split backup ring

installations are more difficult to protect against mechanical damage and can be subject to tidal flows, cross currents, and wave actions.

Even when water filled, PE pipe is less dense than water. Submerged installations require permanent ballast attached around the pipe to ensure submergence and a stable installation once submerged. Ballast design is dependent on the type of installation, and whether cross currents, tidal flows, or wave actions are encountered.

Marine installation requires an appropriate staging site or sites where pipes are fused into long strings and ballast is installed. Bulkheads are attached to pipe ends to prevent the premature entry of water. Staging sites are usually on the shore, but barges are occasionally used for attaching ballast to floating pipelines prior to submergence. Launching typically employs launching slides, ramps, or rails that extend into the water and may be used for attaching ballast to the pipe. As the ballasted pipeline is launched, it is important to control bending and the sinking rate to ensure against kinking. Launching requires appropriate lifting and moving equipment as described in Chapter 7. Flanges and bulkheads should not be used as attachment points for pulling the pipeline.

Where the pipeline is launched into a body of water that is subject to tides, currents, or high wave actions, it may be necessary to control the pipeline with temporary guide cables, anchors, or pilings.

For a submerged pipeline, the ballasted pipeline is floated into position and sunk. Sinking is usually accomplished by slowly filling the pipeline with water from the shore end. At the opposite end of the pipeline, the end is fitted with a bulkhead and a control valve and held above the surface. Sinking rate is controlled by slowly releasing air through the control valve. Sinking must be carefully controlled to ensure against kinking. During sinking, bending should never be tighter than the limits listed in Table 8-7. Once sinking has commenced, it should continue without stoppage at a sinking rate of 800–1,500 feet per minute until the final position is reached[8].

Once submerged, the pipeline is fixed in position. Trenched or on-the-bed installations are typically backfilled with granular material and frequently covered with riprap, large stones, or slabs of broken concrete to protect against backfill erosion. To protect the pipe against additional loads, additional ballast, embedment, and protective cover such as riprap may be required at transition zones such as land-marine areas and at significant underwater grade changes. To keep the pipeline from rolling, exposed on-the-bed installations require ballast weights that have a flat bottom and are bottom heavy.

Subsurface, above-bed pipelines are usually clamped into pillars having contoured saddles. An elastomeric sheet must be wrapped around the pipe to protect the pipe against chafing. Pillars are spaced so that buoyant deflection of the pipe is within design limits. Subsurface pillar supported pipelines may use temporary ballast that is used to sink the pipeline but is then removed after the pipeline is clamped in the pillars.

Occasionally, a PE pipe is not submerged but is floated at or above the surface, and in such cases, ballast may be floats that are strapped to the pipe. Floating pipelines are usually temporary.

Lastly, at land-marine transition areas, the pipe must be protected against shear and bending loads. See Cold (Field) Bending and Controlling Shear and Bending Loads at Transition Areas in this chapter.

Submergence Weighting

A body submerged in a liquid displaces liquid equal to its volume. If the body weighs more than the weight of the liquid volume displaced, it will sink. If it weighs less, it will float. PE materials are lighter than water, and the pipe will float slightly above the surface when filled with water. Submerged pipe must be ballasted to keep it submerged.

Ballast weighting design considers the fluids outside and inside the pipe, the liquid volume displaced, the weight of the displaced liquid volume, the weights of the submerged bodies (pipe, pipe contents, and ballast), and the environmental conditions. Depending on pipe size, individual ballast weights are spaced at intervals along the submerged pipe length.

The following step-by-step method can be used to determine overall ballast weighting for stable submergence of the pipeline and individual ballast weights, but it is not the only means for determining ballast weighting. The method presented requires choosing a ballast weight material and shape.

The method determines the amount of ballast weighting required for stable submergence for 1 ft of pipeline. The designer then chooses the ballast material and the periodic spacing appropriate for the application, and uses the 1-ft value from the method to determine "dry-land" weight for the individual ballast weights that will be attached to the pipeline.

Step 1. Determine volume of external liquid displaced and buoyancy for 1 ft of pipe. This is the uplift that results when the pipe is submerged in its anticipated underwater environment.

$$V_P = \frac{\pi D_o^2}{576} \qquad \text{(Eq 8-5)}$$

$$BP = V_P K_e \omega_{LO} \qquad \text{(Eq 8-6)}$$

Where:

V_P = displaced volume of pipe, ft^3/ft

D_o = pipe outside diameter, in.

B_P = buoyancy of pipe, lb/ft

K_e = underwater environment factor (Table 8-8)

ω_{LO} = specific weight of the liquid outside the pipe, lb/ft^3

The underwater environment factor, K_e, compensates for the effects of tidal flows, uplifts, and currents. Unless neutral buoyancy is desired, K_e should be greater than 1.0.

Table 8-9 presents specific weights for selected fluids. For other liquids, Equation 8-7 may be used to calculate a specific weight when the specific gravity of the liquid is

Table 8-8 Underwater environment factor, K_e

Underwater Environment	K_e
Neutral buoyancy	1.0
Lakes, ponds, slow moving streams or rivers, low currents, and tidal actions	1.3
Significant stream or rover currents, or tidal flows	1.5
Extreme currents, tidal flows, and wave actions at land-shore transitions	1.7

Table 8-9 Specific gravities and specific weights at 60°F (15°C)

Liquid	*Specific Gravity,* S_L	Specific Weight, $_L$
Freshwater	1.00	62.4
Seawater	1.026	64.0
Gasses	0.00	0.00

known. For this discussion, gasses (air, nitrogen, carbon dioxide, etc.) in the pipe have a specific gravity of zero relative to water.

$$\omega_L = 62.4\, S_L \quad \text{(Eq 8-7)}$$

Where:

ω_L = specific weight of liquid

S_L = specific gravity of liquid

Step 2. Determine negative buoyancy for 1 ft of pipe (pipe weight and pipe contents weight). This is negative lift from the weight of 1 ft of pipe and its contents.

$$V_B = \frac{\pi D_j^2}{576} \quad \text{(Eq 8-8)}$$

$$B_N = w_P + (V_B \omega_{LI}) \quad \text{(Eq 8-9)}$$

Where:

V_B = pipe bore volume, ft^3/ft

D_j = pipe average inside diameter, in. (Tables 3-1 and 3-2)

B_N = negative buoyancy, lb/ft

w_P = pipe weight, lb/ft (Tables 3-1, 3-2, and 8-10)

ω_{LI} = specific weight of the liquid inside the pipe*, lb/ft^3

* ω_{LI} – In cases where air may be entrained in water or entrapped at high points in a water pipeline, a reduced value for the specific gravity of water may be used to estimate the specific weight of the liquid inside the pipe. For example, if water in the

Table 8-10 Pipe weight conversion factors

Multiply	By	To Obtain
PE 3408 pipe weight, lb/ft	0.986	PE 2406 pipe weight, lb/ft

pipeline is assumed to contain 2 percent entrained air, the estimated specific gravity of the water-entrained air mixture would be 1.00 – 0.02 or 0.98. If for example entrapped air accumulates in an area and occupies 25 percent of the pipeline volume, the estimated specific gravity at that location would be 1.00 – 0.25 or 0.75. (A significant accumulation of entrapped air can result in increased buoyancy and loss of stable submergence. Where an accumulation of entrapped air may be problematic, periodic maintenance by forcing a soft pig through the line to flush air from the system may be appropriate.)

Step 3. Determine the weight per foot of the submerged ballast. This is the amount of negative lift that must be added to 1 ft of pipe and its contents to counteract buoyant uplift.

$$W_{BS} = B_P - B_N \quad \text{(Eq 8-10)}$$

Where:

W_{BS} = weight of submerged ballast, lb/ft

Step 4. For the chosen ballast weight material, the required dry land weight should be determined. The submerged ballast weight will also displace its volume of liquid. The designer must choose an appropriate ballast weight material (concrete, steel, etc.) and an appropriate weight spacing (Table 8-11). Ballast weights are usually spaced to avoid excessive pipe bending stresses during and after installation. Once the ballast weight material and weight spacing are chosen, the volume of ballast material required to produce the desired submerged weight can be determined and then the dry-land weight of the individual ballast weight. Eq 8-11 is used to calculate the dry-land weight of the individual ballast weights that are attached to the pipe at the chosen span.

$$W_{BD} = \frac{L_{BS} \, W_{BS} \, \omega_B}{(\omega_B - \omega_{LO})} \quad \text{(Eq 8-11)}$$

Where:

W_{BD} = weight of dry ballast, lb

L_{BS} = ballast weight spacing, ft

ω_B = specific weight of ballast material, lb/ft^3

Weight shapes. Ballast weights can be produced from a variety of materials and shapes. Permanent ballast weights are typically produced from reinforced concrete that is formed in molds to the desired size and shape.

Table 8-11 Approximate ballast weight spacing*

Nominal Pipe Diameter		Approximate Span	
in.	*mm*	*ft*	*m*
Up to 12	Up to 300	5 to 10	1.5 to 3
>12 to 24	>300 to 600	7.5 to 15	2.3 to 4.6
>24 to 63	>600 to 1,600	10 to 20	3.0 to 6.1

* Because of the external bead at a butt fusion, ballast weights should not be attached to the pipe such that the weight clamps around the pipe over a butt fusion joint. Butt fusion locations along the pipeline will vary depending on the pipe length shipped from the manufacturing plant. Typical shipping lengths are 40, 45, or 50 feet. The chosen weight spacing should take the probable locations of butt fusions into account, and installation specifications should include instructions to avoid installing a weight over a butt fusion or to remove the external butt fusion bead if installing a ballast weight over a butt fusion cannot be avoided.

Ballast shape is dependent on the application. For example, cylindrical weights can be suitable for a trenched and backfilled installation, where embedment secures the pipe against rolling. Alternatively, an exposed pipeline on the bed would require bottom-heavy rectangular weights that would prevent rolling from currents and wave actions. Temporary ballast may be weights at the ends of straps that are draped over the pipeline for submergence and removed when the pipeline is secured in place. In some cases where submergence is shallow, inverted U-shaped "set-on" weights can be placed over the pipeline to sink it in place.

Submergence weights are frequently made of reinforced concrete, which allows considerable flexibility of shape design. Weights are typically formed in two or more sections that clamp around the pipe over an elastomeric padding material. There should be clearance between the sections, so when clamped onto the pipe, the ballast weights do not slide along the pipe. Figures 8-14 and 8-15 illustrate flat bottom, bottom heavy weights for on-the-bed submerged pipelines. This design prevents rolling from cross-current conditions. Fasteners securing the weight sections together must be resistant to the marine environment.

Floating Pipelines

In some cases, it may be appropriate for a pipeline to float at or above the surface. For example, a pipeline may extend out into a reservoir and connect to a floating suction or discharge structure, or a temporary pipeline may be needed to reduce a reservoir level. Flexible PE pipe can be an excellent candidate for such applications. Floating pipelines may be either supported above the surface on individual support floats (Figure 8-16) or floated at the surface by strapping floats lengthwise along the pipeline (Figure 8-17). When the pipeline is supported above the surface, the floats must support their own weight and the weight of the pipeline and its contents. When floated at the surface, the displacement of the pipeline in the water reduces flotation requirements.

Figure 8-16 illustrates floats where the pipeline is cradled in a support structure above the float, and only the float is in the water. A single float is illustrated, but floats may be in pairs or multiples at the discretion of the designer. Figure 8-17 illustrates flotation at the surface where the floats *and the pipeline* are in the water. In the illustration, twin parallel float pipes are strapped to the pipeline.

Floats are usually pipe lengths that are capped on the ends. For its durability and light weight, thin-walled PE pipe can be used as floats. Floats can be filled with lightweight foam so that a punctured or damaged float cannot fill with water and sink.

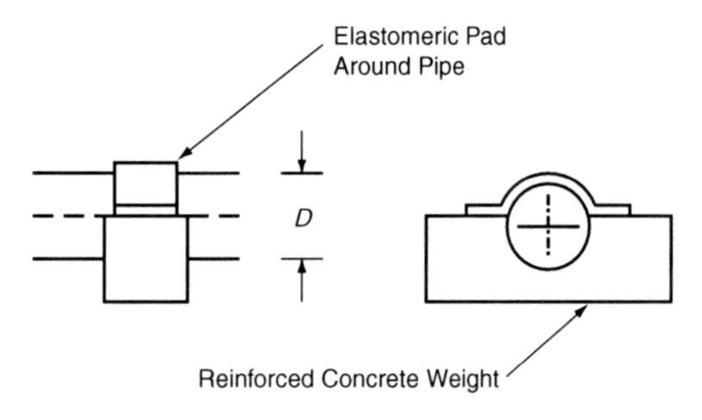

Figure 8-14 Concrete weight

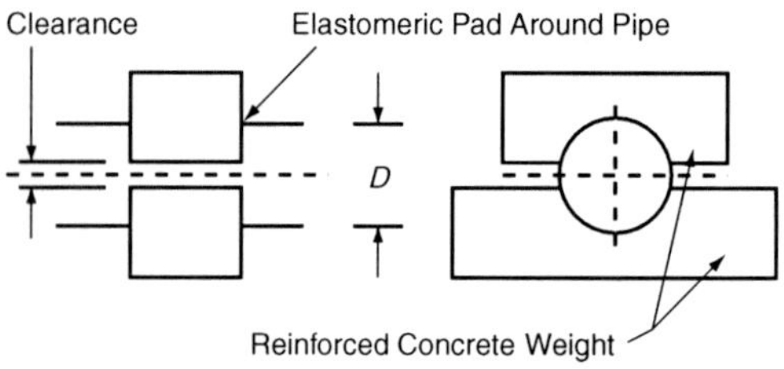

Figure 8-15 Concrete weight

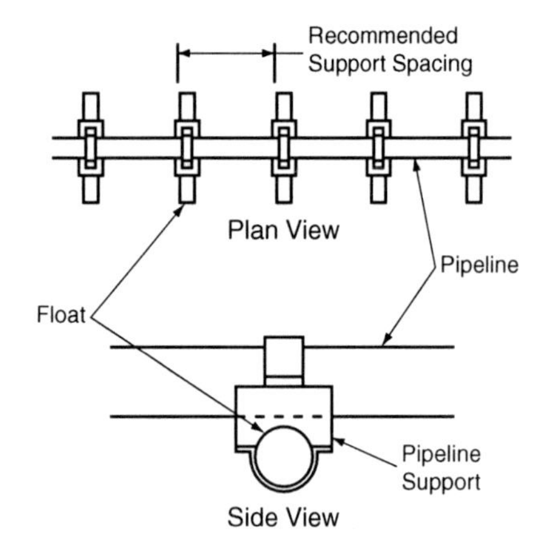

Figure 8-16 Flotation above the surface

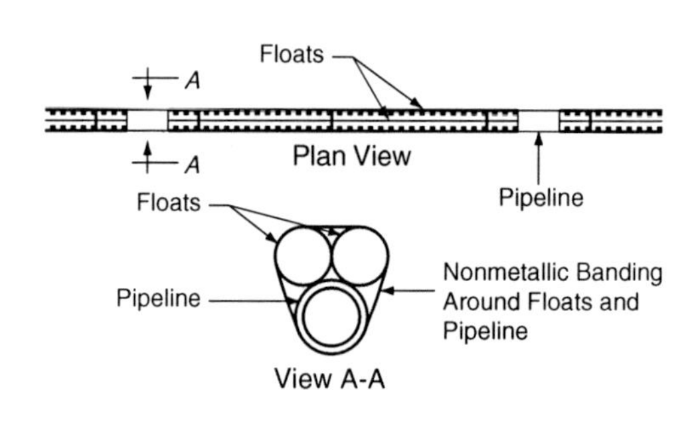

Figure 8-17 Flotation at the surface

Float sizing. Float sizing is an iterative process because the float must support itself as well as the load on the float. The first step is to determine the load and choose an initial float size. The procedure that follows is useful for determining float design for flotation above the surface or at the surface. The primary difference is that a pipeline that is supported at the surface requires less support because of the pipeline's displacement in the water.

Step 1: Load determination. The supported load is the weight of the pipeline and its contents plus the weight of the float and the structure for attaching the float to the pipeline. If the float is foam-filled, the weight of the foam must also be included.

For a Figure 8-16 floating pipeline,

$$W = L_{SP}(w_P + w_{LI}) + w_S + L_F(w_F + w_M) \qquad \text{(Eq 8-12)}$$

For a Figure 8-17 floating pipeline, the pipeline and its contents are buoyed by their submergence. Therefore, the supported load, W, is determined using Eq 8-5 through 8-9, and $K_e = 1.0$ to determine B_N, and Eq 8-12 can be rewritten as:

$$W = L_{SP}(B_N + w_{LI}) + w_S + L_F(w_F + w_M) \qquad \text{(Eq 8-13)}$$

Where terms are previously defined and

W = supported load, lb

B_N = negative buoyancy, lb/ft
L_{SP} = length of supported pipeline, ft*
w_P = weight of pipeline, lb/ft
w_{LI} = weight of pipeline contents, lb/ft
w_S = weight of float attachment structure, lb
L_F = length of float, ft
w_F = float weight,† lb/ft (Table 8-12)
w_M = weight of foam fill,† lb/ft

$$W_M = V_F M_M \quad \text{(Eq 8-14)}$$

V_F = float internal volume, ft^3/ft (Table 8-12)
M_M = density of foam fill, lb/ft^3 (Thermoplastic foams typically weigh 2 to 3 lb/ft^3.)

* When a pipeline is supported above the surface, L_{SP} is the distance (spacing) between the support floats. When a pipeline is supported at the surface, L_{SP} is the length of pipeline supported by longitudinal floats plus any longitudinal unsupported pipeline distance between floats. For example, if a pipeline is supported by equally spaced, parallel 40-ft long floats, and the end-to-end gap between floats is 20 ft, L_{SP} is the supported length plus the unsupported length; that is, L_{SP} = 40 ft + 20 ft = 60 ft.

† For a pipeline supported at the surface by twin, parallel floats strapped to the pipeline, as illustrated in Figure 8-17, each float supports half of the load, but the float weight (w_F) must be the weight of both floats plus any internal float foam fill.

Support spacing. When a pipeline is supported above the surface, float spacing must be determined so that the load on individual floats may be determined. (See Figure 8-18.) Support spacing depends on the allowable deflection between supports, which in turn depends on the pipeline, the fluid within it, and the service temperature. In general, the allowable long-term deflection between supports should not exceed 1 in. to avoid high spots at the float support points where entrapped air could collect. Support spacing may be determined from Equation 8-15.

$$L_S = \frac{1}{12}\left(\sqrt[4]{\frac{384 E I y_S}{\frac{5}{12}(w_p + w_{LI})}}\right) \quad \text{(Eq 8-15)}$$

Where:

L_S = distance between floats, ft
E = long-term modulus for the service temperature, $lb/in.^2$ (Table 5-6)
I = moment of inertia, $in.^4$
y_S = deflection between floats, in.
w_p = weight of pipeline, lb/ft
w_{LI} = weight of pipeline contents, lb/ft

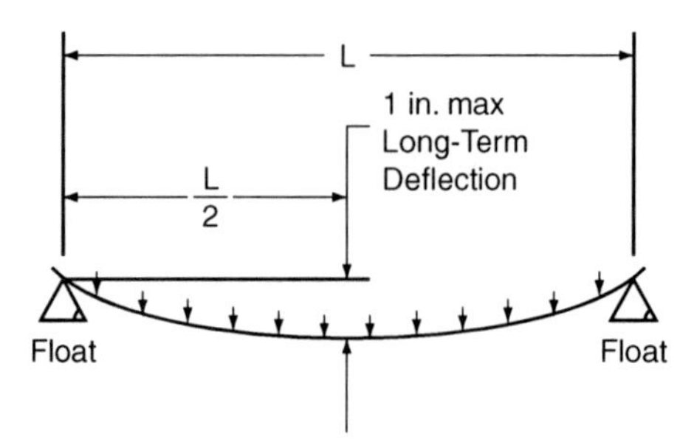

Figure 8-18 Deflection between floats

$$I = \frac{\pi(D_o^{\,4} - D^4)}{64} \qquad \text{(Eq 8-16)}$$

D_o = pipeline outside diameter, in. (Table 8-12)

D = pipeline inside diameter, in. (Tables 3-1 and 3-2)

Step 2: Float submergence percentage. The ability of a cylindrical float to support a load depends on how much of the float is submerged. The percent submergence is the percent of the float that is below the water level as illustrated in Figure 8-19.

$$\%\ \text{submergence} = 100\frac{h}{d} \qquad \text{(Eq 8-17)}$$

Where:

h = float submergence below water level, in.

d = float diameter, in. (Table 8-12)

The designer should choose an appropriate percent submergence by determining an appropriate reserve load capacity for the float. (See Table 8-13 and Eq 8-17.) If, for example, the load, P, causes the float to be half submerged, it will require twice the load, $2P$, to fully submerge the float, and if the float is fully submerged, additional load will sink the float. Additionally, if the percent submergence is too high, centrally loaded flexible PE pipe support floats such as those illustrated in Figure 8-16 may deflect and submerge more deeply at the load center. Longitudinal floats, such as those illustrated in Figure 8-17, are more uniformly loaded along their length.

Step 3: Float support capacity. Determine the float buoyancy, B, from Table 8-12 for the initial float size. Then determine the submergence factor, f_S, from Table 8-13.

Determine the load supporting capacity of the float, W_F.

$$W_F = L_F f_S B \qquad \text{(Eq 8-18)}$$

Where:

W_F = float load supporting capacity, lb

f_S = submergence factor from Table 8-13

B = float buoyancy from Table 8-12

L_F = length of float, ft

Table 8-12 Polyethylene float properties*

Nominal Size	Float Outside Diameter, d, *in.*	Float Weight, W_F, *lb/ft*	Float Buoyancy, B, *lb/ft*	Internal Volume, V_F, ft^3/ft
4	4.500	0.83	6.9	0.097
6	6.625	1.80	14.9	0.211
8	8.625	3.05	25.3	0.357
10	10.750	4.75	39.3	0.555
12	12.750	6.67	55.3	0.781
14	14.000	8.05	66.7	0.941
16	16.000	10.50	87.1	1.230
18	18.000	13.30	110.0	1.556
20	20.000	16.41	136.0	1.921
22	22.000	19.86	165.0	2.325
24	24.000	23.62	196.0	2.767
26	26.000	27.74	230.0	3.247
28	28.000	32.19	267.0	3.766
30	30.000	36.93	306.0	4.323
32	32.000	42.04	349.0	4.919
34	34.000	47.43	393.0	5.553
36	36.000	53.20	441.0	6.225

* Float properties are based on floats produced from DR 32.5 black, high-density PE pipe material (0.955 g/cm^3 density) and flotation in fresh water having a specific weight of 62.4 lb/ft^3.

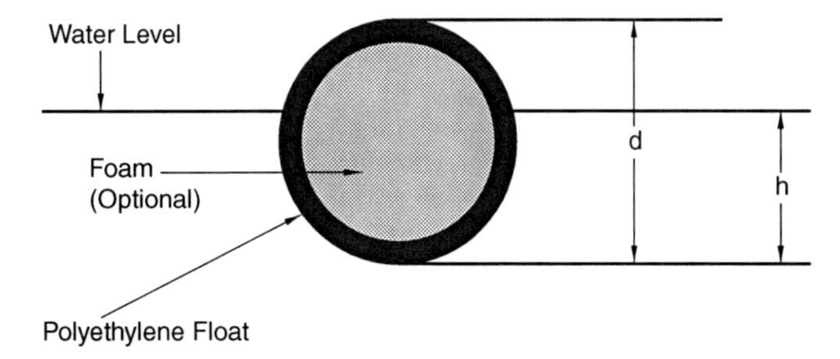

Figure 8-19 Float submergence

Table 8-13 Submergence factor, f_S

Submergence		Submergence		Submergence		Submergence	
Percent	Factor, f_S	Percent	Factor, f_S	Percent	Factor, f_S	Percent	Factor, f_S
5	0.019	30	0.252	55	0.564	80	0.858
10	0.052	35	0.312	60	0.623	85	0.906
15	0.094	40	0.377	65	0.688	90	0.948
20	0.142	45	0.436	70	0.748	95	0.981
25	0.196	50	0.500	75	0.804	100	1.000

Step 4: Compare float support capacity to load. The support capacity of the float must equal or exceed the load it is to support.

$$W_F \geq W \qquad \text{(Eq 8-19)}$$

If the load, W, is greater than the float support capacity, W_F, choose a larger float and repeat Steps 1, 2, and 3. If the float support capacity, W_F, is significantly greater than the load, W, a smaller float may be adequate. The load W applied to the float is also dependent on the number of floats supporting the load. If the total load is supported by two floats, each float supports half the total load or $W/2$.

Step 5: Check actual float submergence. Once the proper float size has been determined, check the actual float submergence.

$$f_{SA} = \frac{W}{B} \qquad \text{(Eq 8-20)}$$

Where:

f_{SA} = actual float submergence factor

The actual float submergence factor, f_{SA}, may be compared to the values in Table 8-13 to determine the approximate percent submergence.

REFERENCES

1. ASTM D2774 *Standard Practice for Underground Installation of Thermoplastic Pressure Piping,* ASTM International, West Conshohocken, PA.
2. ANSI/AWWA C906, *Polyethylene (PE) Pressure Pipe and Fittings, 14 In. (100 mm) Through 63 In. (1,575 mm), for Water Distribution and Transmission,* American Water Works Association, Denver, CO.
3. ASTM D698 *Standard Test Methods for Laboratory Compaction Characteristics of Soil Using Standard Effort* (12,400 ft-lbf/ft^3 (600 kN-m/m^3)), ASTM International, West Conshohocken, PA.
4. ASTM F2164 Standard *Practice for Field Leak Testing of Polyethylene (PE) Pressure Piping Systems Using Hydrostatic Pressure,* ASTM, West Conshohocken, PA.
5. ASTM D2487 Standard *Classification of Soils for Engineering Purposes (Unified Soil Classification System),* ASTM International, West Conshohocken, PA.
6. ASTM D2488 *Standard Practice for Description and Identification of Soils (Visual-Manual Procedure),* ASTM International, West Conshohocken, PA.
7. ANSI/AWWA C901, *Polyethylene (PE) Pressure Pipe and Tubing, ½ In.(13 mm) Through 3 In. (76 mm), for Water Service,* American Water Works Association, Denver, CO.
8. ASTM F1804 *Standard Practice for Determining Allowable Tensile Load for Polyethylene (PE) Gas Pipe During Pull-In Installation,* ASTM International, West Conshohocken, PA.
9. ASTM F1962 *Standard Guide for Use of Maxi-Horizontal Directional Drilling for Placement of Polyethylene Pipe or Conduit Under Obstacles, Including River Crossings,* ASTM International, West Conshohocken, PA.
10. Marine Installations, *PPI Handbook of Polyethylene Piping,* Plastics Pipe Institute, Inc, Washington, DC.

Chapter 9

Hydrotesting and Commissioning

FLUSHING

To prevent damage to valves or other fittings from any foreign material left in the pipeline, the pipe should be thoroughly flushed prior to testing. Flushing can be accomplished by opening and closing hydrants, blow-offs, or drains with flow velocities sufficient to flush the foreign material from the pipeline. A minimum velocity of 3 ft/sec is suggested. The initial flushing should be continued until the discharge appears clean; however, the minimum duration should be based on three changes of pipeline volume.

FILLING

The pipeline should be filled slowly, limiting the flow to low velocities that prevent surges and air entrapment. Air valves at high points should be open to allow air to escape as the water level increases inside the pipeline. If permanent air valves are not required at all high points, the contractor should install temporary valves at these points to expel air during filling. Loosening flanges or connections to bleed air from the system is prohibited. The critical filling rate for pipes with air vents is usually based on 5 to 15 percent of the pipe design flow. For air valves, the filling rate is limited by orifice size and the fact that the seat will blow shut when air passing through the valve reaches sonic velocity. A typical maximum filling rate for a pipe system with 2-in. air valves is 2 ft^3/sec. However, the maximum filling velocity in the pipeline should never exceed the design velocity.

LEAK TESTING

Before pressure is applied, the pipeline section under test should be restrained against movement. Failure during pressurization can be sudden and violent. All persons must be kept away from pressurized lines at all times . For safety purposes, binoculars can be used to inspect exposed joints for leakage.

Leak Test Procedure

Leak tests should be conducted in accordance with ASTM F2164 at a pressure and for a duration of time agreed on between the owner and designer. (See NOTE below.) For PE pipe, a pressure of 1.5 times the design working pressure at the lowest point in the test section is used as the test pressure. The specified test pressure, based on the elevation of the lowest point of the line or section under testing and corrected to the elevation of the test gauge, is applied using a pump connected to the pipe. The test pressure is always taken at the lowest point in elevation along the test section's pipeline profile. However, the pressure applied must not exceed the design pressure of any fitting, pipe, or thrust restraint system used on the pipeline. Any exposed pipe or fittings should be examined carefully during the leak test for damage. Any damaged or defective pipe, fittings, valves, or hydrants discovered during the leak test should be repaired and the test repeated. During the test period, makeup water is added to keep the pressure constant. During the initial expansion phase, the expansion of the PE pipe will be logarithmic. However, after the four-hour period of pressurization at 1.5 times the hydrostatic design stress (HDS), the pipe expansion becomes more linear.

NOTE: The user should be aware that ASTM F2164[1] is a field leak testing procedure, not a pressure test of the system. In PE piping systems, field pressure tests cannot be used to determine system pressure capacity. (See Chapter 4.) Under no circumstances should the total time for initial pressurization and time at test pressure exceed eight hours at 1.5 times the system pressure rating. If the test is not completed because of leakage, equipment failure, or any other reason within this total time, the test section should be depressurized and allowed to "relax" for at least eight hours before starting the next testing sequence.

Acceptance criteria. If the pressure remains steady (within 5 percent of the target value) for one hour, leakage is not indicated. (ASTM F2164 does not specify a makeup water allowance or require monitoring make-up water volume.)

Testing Inside the Trench

If agreed to by the supervising engineer, leak testing with water can be conducted after joining is complete and the pipe has been backfilled or adequately blocked to prevent movement. Before beginning the test, all test equipment valves are checked to ensure that they are open; connections and valves are checked to ensure that they are completely made up and secured; and equipment is checked to ensure that it is fully operational. During the test, two gauges are desirable at the pump to ensure against an erroneous pressure reading that could allow excess pressures that might damage the pipeline. Meters for measuring makeup water and the pressure gauge at the low point are generally furnished and calibrated by the owner.

NOTE: Leak testing using a pressurized gas (pneumatic testing), such as air, is dangerous and is prohibited. Always use an environmentally safe liquid, such as clean water, for leak testing.

The water, pipe, and soil should be allowed to thermally stabilize. Usually the pipeline is filled, the air vented, and the filled pipeline allowed to sit overnight (in above freezing weather) to thermally stabilize. The time required for thermal equalization will depend on the fill water temperature, the size of pipeline, and weather conditions.

NOTE: Ensure that all air is removed from the system before applying pressure to the pipeline system.

Table 9-1 Standard pressure class

	Material	
Dimension Ratio (DR/IDR)	PE 3408	PE 2406
32.5/30.5	51	40
26/24	64	50
21/19	80	63
17/15	100	78
15.5/13.5	110	86
13.5/11.5	128	100
11/9	160	125
9/7	200	156
7.3/5.3	254	198
6.3/4.3	302	236

NOTE: Table 9-1 values are for a hydrostatic design basis of 1,600 psi for PE 3408 and 1,250 psi for PE 2406 at 73°F, and a 0.50 design factor. Per ANSI/AWWA C906[2], the pressure class applies to operating temperatures through 80°F.

Tests should be performed on reasonable lengths of pipelines. Longer lengths make leak detection more difficult. Generally, pipelines longer than 5,000 ft are tested in shorter sections. Test sections are sealed with full pressure rated end closures such as metal blind-flanges bolted to PE flange adapters with steel backup rings. Blocking to hold the pipe in place during testing is required. Pressure leak testing against closed valves is not permitted because of the potential for leakage of entrapped air and damage to the valve.

The test section should be subjected to a maximum hydrostatic test pressure of 1.5 times the rated pressure of the pipe (1.5 times the standard pressure class as obtained from Table 9-1) or the maximum pressure rating of the lowest pressure rated component in the test section, whichever is less, for a maximum period of three hours. The pipe can be maintained at the "test pressure" by the periodic addition of make-up water. The "test pressure" is monitored at the lowest elevation point of the test section. The test pressure should never exceed the maximum pressure rating for any fitting in the pipeline system. When, in the opinion of the supervising engineer, local conditions require that the trenches be backfilled immediately after the pipe has been laid, the leak test may commence after the backfilling has been completed but not before seven days after the last concrete bearing pad or thrust block has been cast.

Testing Outside the Trench

When slip lining or directional drilling operations call for a leak test prior to installation of the pipe and if agreed to by the supervising engineer, leak testing with water can be conducted after joining is complete and before laying the pipe in the trench.

NOTE: Testing of PE pipe outside of the trench prior to installation is dangerous. Ensure that the pipeline has been blocked to prevent movement in case of joint rupture and that there are no persons near the pipe while the pipe is pressurized.

All of the procedures outlined in the previous section should be followed for testing inside the trench.

As leaks are determined by visual examination (binoculars may be used to ensure personal safety), it is not necessary to calibrate the makeup water for the initial stretching of the pipe. All joints should be examined for leakage.

NOTE: Correctly made fusion joints do not leak. If leakage is observed at a fusion joint, complete rupture may be imminent. Immediately move all persons away from the joint and depressurize the pipeline.

Any joints showing leakage must be repaired and the system retested. Faulty fusion joints must be removed and remade.

Leakage

Because joints for PE pipe are fused together, the amount of leakage should be zero. The critical leakage rate for the pipeline system is usually specified in the contract documents.

If any tests show leakage greater than that allowed, the installer is responsible for locating and repairing the leak and retesting until the test result is within acceptance criteria. Retesting can be performed after depressurizing the pipeline and allowing the pipeline to "relax" for at least eight hours. All visible leaks should be repaired.

NOTE: Never attempt to repair leaks while the system is under pressure. Always depressurize the system before repairing leaks.

RECORDS

The test records should include (as a minimum) the following documentation:

1. Test medium (normally water)
2. Test pressure
3. Test duration
4. Test data
5. Pressure recording chart or pressure log
6. Pressure versus makeup water added chart
7. Pressure at high and low elevations
8. Elevation at point test pressure is measured
9. Ambient temperature and weather conditions
10. Pipe and valve manufacturers
11. Pipe specifications and/or standards (ASTM, AWWA, etc.)
12. Description of the test section length, location, and components
13. Description of any leaks, failures, and their repair/disposition. Person or contractor conducting the test
14. Test times and dates

DISINFECTION

Newly installed potable water pipelines require disinfection in accordance with ANSI/AWWA C651[3]. The disinfection should take place after the initial flushing and after the completion of the pressure testing. Provisions should be made to avoid contamination of existing mains by cross-connection during flushing, testing, and disinfection of newly installed pipelines.

Prolonged exposure to certain disinfection chemicals in high or extreme concentrations may be damaging to the inside surface of PE pipe and is to be avoided. Disinfecting solutions containing chlorine should not exceed 12 percent active chlorine. As soon as the normal pipe disinfection period is over, the disinfection solution should be purged and/or neutralized from the pipeline, and the pipeline should be filled with fresh, clean water. Purging applies to distribution mains as well as each service line and service connection.

COMMISSIONING

The commissioning of new or repaired pipeline is normally carried out in the following sequence of events:

1. Cleaning and/or pigging of the pipeline
2. Water filling and disinfection sterilization
3. Flushing, purging, and/or neutralization
4. Refilling the pipeline
5. Bacteriological sampling and testing
6. Certifying and accepting
7. Initiating pipeline into service

This sequence is basic to commissioning PE pipelines but may be adapted to meet particular conditions or project requirements.

REFERENCES

1. ASTM F2164, *Standard Practice for Field Leak Testing of Polyethylene (PE) Piping Systems Using Hydrostatic Pressure*. ASTM International West Conshohocken, PA.
2. ANSI/AWWA C906, *Polyethylene (PE) Pressure Pipe and Fittings, 4 In. (100 mm) Through 63 In. (1,575 mm), for Water Distribution,* American Water Works Association, Denver, CO.
3. ANSI/AWWA C651, *Disinfecting Water Mains,* American Water Works Association, Denver, CO.

This page intentionally blank.

Chapter 10

Maintenance and Repairs

DISINFECTING WATER MAINS

The applicable procedures for disinfecting new and repaired potable water mains are presented in ANSI/AWWA C651[1], *Disinfecting Water Mains.* All repairs and new installations should be disinfected prior to using the line to transport water for the general public.

The installer is cautioned to dispose of the chlorinated test and disinfection medium in a safe and environmentally acceptable fashion. The chlorine residual is hazardous to fish and animals. All disposals should be in accordance with local, state, or federal code.

CLEANING

New installations or lines after repair may be cleaned via the water-jet process or forcing a soft pig through the line. Water-jet cleaning is available from commercial services. Cleaning usually employs high-pressure water sprayed from a nozzle that is drawn through the pipe system with a cable.

Pigging involves forcing a resilient plastic plug (soft pig) through the pipeline. Usually, hydrostatic or pneumatic pressure is applied from behind to move the pig down the pipeline. Soft pigs must be used with PE pipe. Scraping finger-type pigs will damage the PE piping material and must not be used. Commercial pigging services are available if line pigging is required.

MAINTENANCE

A good maintenance program for PE water distribution and main piping systems will include the same elements utilized for other available piping materials. It is recommended that the system operator have

1. A means for locating the pipe
2. Records of system performance including:

    - Unbillable water loss (PE piping systems may outperform other piping materials because of zero-leakage heat fusion joints)

- Tests for system efficiency including flow coefficient
- Tools required for installation and repair including squeeze-off tools and heat fusion equipment
- Repairs, downtime, and causes

Pipe Locating

Because of PE's material makeup, it is not locatable using detection systems common to metallic piping materials. Therefore, detection systems require installing a tracer wire or plastic coated metallic tape in the trench during PE pipe installation. Inductive electronic pipe tracer systems are available that inductively detect the metallic tracer wire or plastic coated metallic tape from the surface. Detection of pipelines at greater depths can be successfully achieved through conductive detection where the detector is physically connected to the tracer wire or tape. Acoustic sounding devices can also be used.

Tools

PE pipe is suitable for squeeze-off; however, the manufacturer should be consulted for applicability and procedures. Water flow control by squeeze-off is available for PE piping systems. Because of its long successful history in use as gas distribution piping, several ASTM standards are available for proper squeeze tools and application technique. Squeeze-off is generally limited to pressure pipe up to 18-in. IPS and 100-psi internal pressure.

The squeeze-off tool should be designed for use on PE pipe, and it is recommended that the tool be designed in accordance with ASTM F1563[2]. This will ensure that the tool was designed to minimize the potential for material damage. Closing and opening rates are key elements to performing squeeze-off. It is necessary to close slowly and release slowly, with slow release being more important. PE pipe or tubing is never subjected to squeeze-off more than once in the same place.

Butt fusion equipment is needed to join PE pipe and fittings. Butt fusion is the best method of joining PE pipe and when correctly made, provides leak-free joints. The reliability and ease of use is well documented. Equipment can be purchased or rented.

Electrofusion saddles are used to make repairs and attach new service connections. Electrofusion couplings are used for final tie-ins and to connect MJ adapters and flange adapters to PE pipe when butt fusion is inconvenient. Electrofusion is a favored method for making in-the-trench connections and repairs. Equipment can be rented or purchased.

Saddle fusion is used to attach saddles for repairs, services, and branch connections to mains. Saddle fusion equipment is available for rental or purchase.

REPAIRS

Repairing PE pipe is similar to repairing ductile iron and PVC water pipe. Mechanical couplings or fusion methods can be used. The first step in making a repair is determining the problem. In this section, puncture, rupture, and saddle repairs are discussed.

Puncture

A puncture can cause a small hole in the pipe. The most common source of a puncture is the tooth of a backhoe bucket. The tooth of a backhoe striking PE pipe often causes a small hole up to 3 in. across (Figure 10-1). Punctures this size and smaller can be

Courtesy ISCO Industries, LLC

Figure 10-1 Damage to PE pipe by backhoe bucket

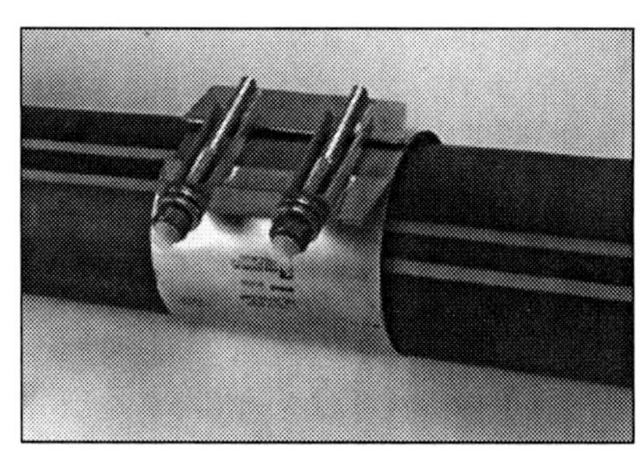

Courtesy ISCO Industries, LLC

Figure 10-2 Wrap-around repair sleeve used to repair small puncture holes in PE pipe

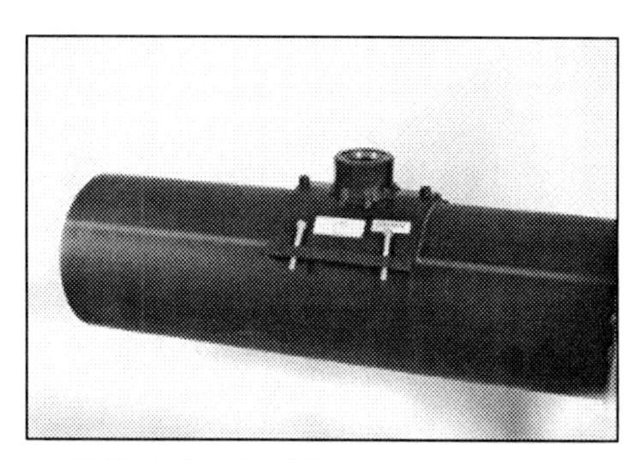

Courtesy ISCO Industries, LLC

Figure 10-3 Electrofusion saddles are available for repairs

repaired using a standard repair sleeve. (Larger punctures may require replacing the damaged section. See the Rupture section.) The methods of repair are:

- Mechanical repair sleeve
- Fusion saddle

Repair sleeves are made to repair water pipe, and the same repair sleeves used to repair PVC pipe can be used to repair PE. The gasket material used in PVC repair sleeves can normally be used with PE pipe. The manufacturer of the repair sleeves should be consulted for the best gasket for PE pipe. Repair sleeves can be installed under wet conditions. Dirt or mud should be removed, and the sleeve installed in accordance with the manufacturer's instructions.

In water systems with relatively constant pressure and temperature, repair sleeves work very well for long periods (Figure 10-2). Where the pressure and/or temperature of the water change greatly, some leaks may occur because of pipe movement relative to the seal. This is a temporary repair, and a permanent repair needs to be considered.

Fusion and electrofusion saddles can also be used to repair punctures (Figure 10-3). To use these saddles, a clean dry surface is required. The saddle is centered over the puncture, fused to the pipe, and (if necessary) the outlet is plugged or capped. Some fusion repair saddles are made with permanently sealed outlets.

Electrofusion and sidewall fusion saddles can repair punctures up to about 2 in. across. Electrofusion and sidewall saddle repairs work well in both constant and variable temperature and pressure conditions.

Rupture

When a backhoe or other outside force ruptures, severs, or severely damages PE pipe, the damaged section must be cutout and replaced. Restrained mechanical, electrofusion couplings or flange adapters can be used for these repairs. Self-restraining mechanical couplings usually have "teeth" that bite into the PE pipe surface to restrain against pullout separation. Figure 10-4 illustrates that two couplings and a piece of PE pipe are needed for repair. Self-restrained mechanical coupling repairs can be made in wet, muddy conditions. The pipe must be depressurized.

A second type of mechanical coupling is a bolted, sleeve-type coupling. Most manufacturers of these couplings recommend the use of insert stiffeners to reinforce the pipe against OD compression from the coupling. Pipe surfaces must be clean, and the pipe must be depressurized. Figures 10-5 and 10-6 show couplings and insert stiffeners.

Electrofusion couplings can also be used to make rupture repairs. As illustrated in Figure 10-7, two electrofusion couplings are used in place of mechanical couplings. When electrofusion couplings are used, the pipe must be clean, dry, and depressurized. Stiffeners are not required for electrofusion couplings.

A flanged spool may also be prepared and installed to replace the damaged section. The damaged section is cut out; flange adapters with backup rings are connected to the exposed pipe ends; and a flanged spool that spans the gap is prepared and installed. Figure 10-8 illustrates this replacement method.

For smaller pipes that can be deflected within the repair excavation, the damaged section can be cutout and a new section butt fused to one of the exposed ends. At the opposite end, a mechanical coupling or electrofusion coupling is installed to complete the repair. Figure 10-9 illustrates this replacement method.

Wrap around couplings as shown in Figure 10-2 are not suitable for ruptured pipe repair because they provide no resistance to end movement.

Saddle

The third type of repair is to repair a damaged saddle. Again, the most common cause of a problem is when a backhoe working near the service hits the saddle or the service line. Figure 10-10 illustrates the problem.

A threaded outlet repair saddle can be used to solve this problem. The saddle outlet is threaded for use with a corp stop (Figure 10-11).

When a service line needs repair, small diameter compression couplings are available. Figures 10-12 through 10-15 show various compression couplings. An insert stiffener must be installed in the end of the PE pipe or tubing whenever compression couplings are used.

These couplings and adapters make it easy to use and repair PE service tubing in water systems. Fittings with male and female threads, other styles, and for connecting PE to copper tubing are also available.

NEW SERVICE CONNECTIONS

New service connections can be made on PE pipe with mechanical saddles, electrofusion saddles, and sidewall fusion saddles.

Mechanical Saddles

Mechanical saddles are similar to saddles available for PVC and ductile iron. Manufacturers of these saddles indicate that they can be used with a water temperature from 35°F (1.7°C) to 85°F (29.4°C).

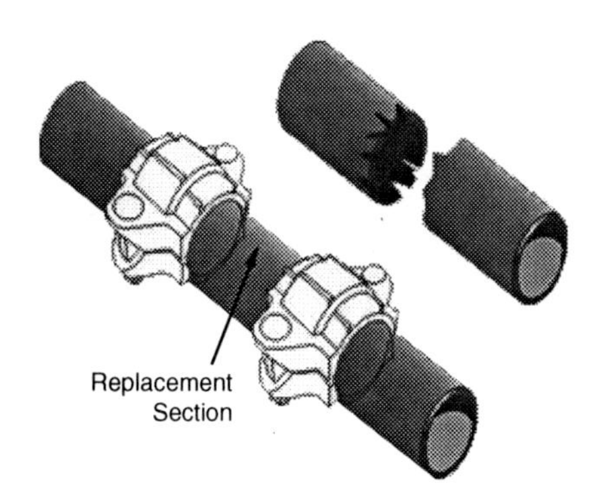

Courtesy ISCO Industries, LLC

Figure 10-4 Repair using self-restraining mechanical couplings

Courtesy JCM Industries

Figure 10-5 Bolted, sleeve-type coupling

Courtesy JCM Industries

Figure 10-6 Insert stiffener

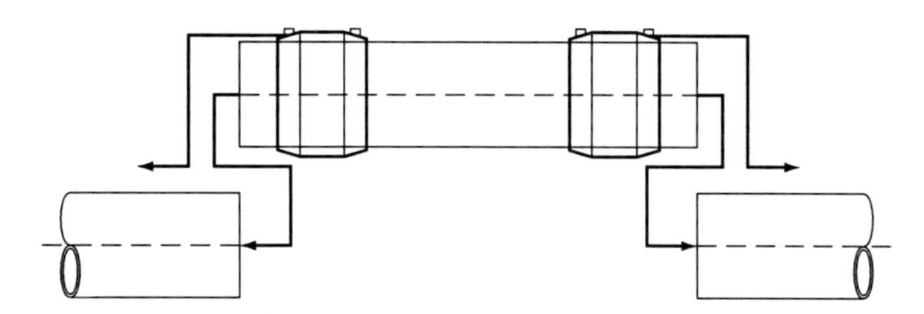

Courtesy Performance Pipe

Figure 10-7 Repair using electrofusion couplings

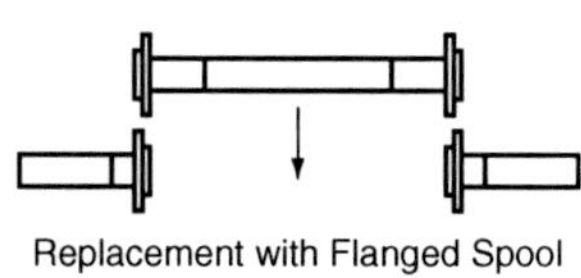

Courtesy Performance Pipe

Figure 10-8 Section replacement using flanged spool

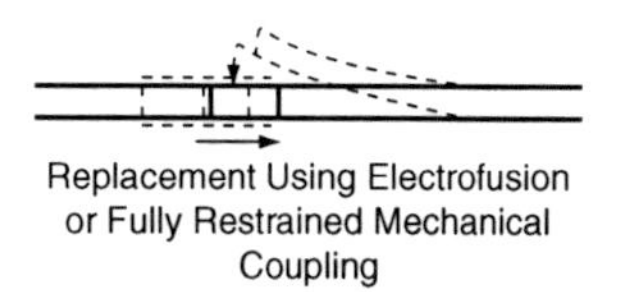

Figure 10-9 Section replacement by deflecting smaller pipes

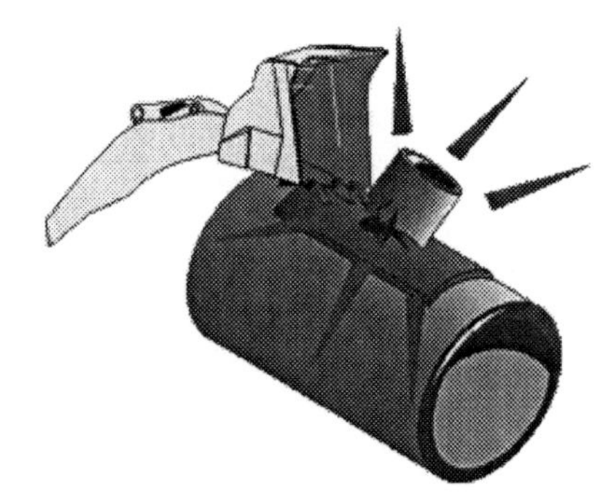

Courtesy Romac Industries, Inc.

Figure 10-10 Damage to saddle by backhoe

Courtesy CEPEX USA, Inc.

Figure 10-11 Threaded outlet repair saddle

Courtesy CEPEX USA, Inc.

Figure 10-12 90° ell compression fitting

Courtesy CEPEX USA, Inc.

Figure 10-13 Straight coupling

Courtesy CEPEX USA, Inc.

Figure 10-14 Male adapter

Courtesy ISCO Industries, LLC

Figure 10-15 Reducing coupling

A mechanical saddle can be installed under a wide range of conditions using common hand tools. Installation is similar to techniques used to install saddles on other piping materials. Figure 10-16 shows the bolts on the saddle being tightened. Figure 10-17 shows a corp stop attached to the saddle. Mechanical saddles should have wide straps (not U-bolts) to distribute compressive forces and must be installed in accordance with the manufacturer's instructions.

Electrofusion and Sidewall Fusion Saddles

Electrofusion and sidewall fusion saddles are also used for service connections (Figure 10-18). These saddles are attached by electrofusion and sidewall fusion. These devices have a long history of successful use in the natural gas industry and are the

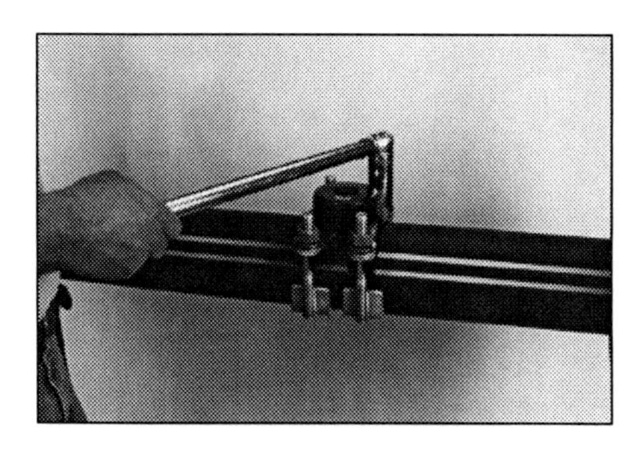

Courtesy ISCO Industries, LLC

Figure 10-16 Attachment of mechanical saddle

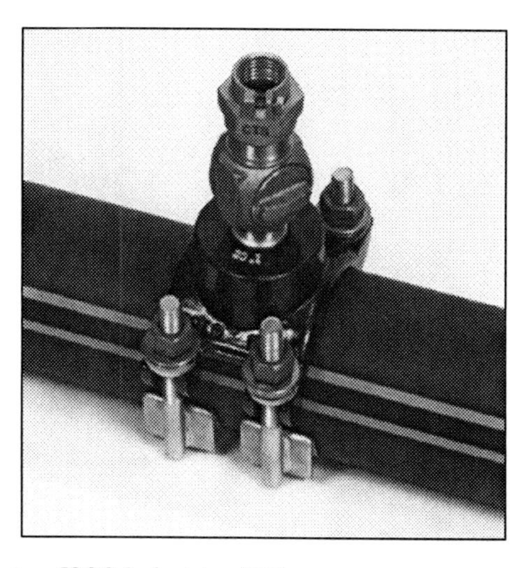

Courtesy ISCO Industries, LLC

Figure 10-17 Corp stop attached to mechanical saddle

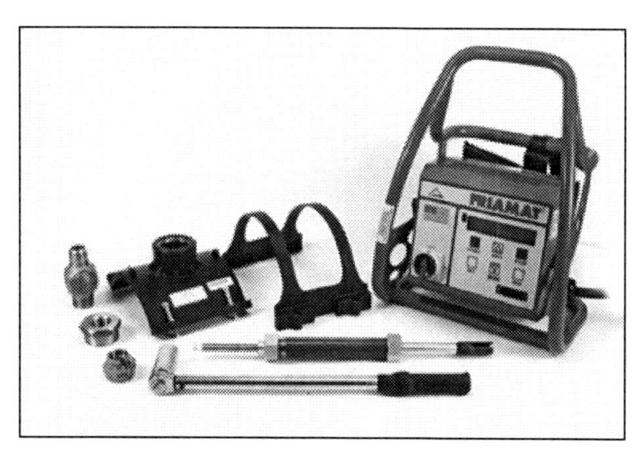

Courtesy ISCO Industries, LLC

Figure 10-18 Electrofusion saddle, processor, corp stop, adapters, wrench, and cutter for hot tap

preferred means of connecting to PE piping because correctly made electrofusion and sidewall fusion joints do not leak.

The pipe and saddle must be clean and dry to attach an electrofusion saddle. Figure 10-19 shows a saddle that has been fused on the main pipe and fitted with corp stop and hot tap cutter for piercing the main. Hot taps are easy to make with PE pipe and saddles.

Sidewall fusion can be used to attach saddles to existing PE pipelines. Figure 10-20 shows a 6-in. saddle for use on 8-in. PE pipe.

A special "curved heater" and combination butt fusion/sidewall fusion machine are used to fuse saddles on PE pipe. After the saddle is attached, a tap can be made with a special cutter. Sidewall fusion should be performed in accordance with manufacturer's instructions.

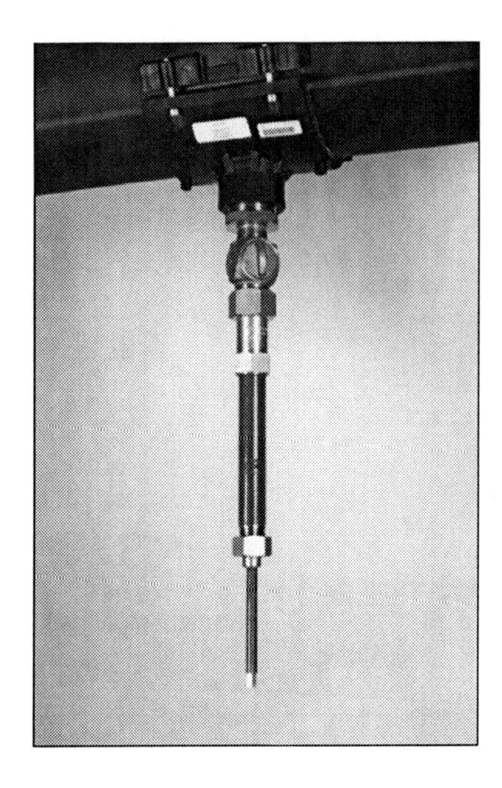

Courtesy ISCO Industries, LLC

Figure 10-19 Electrofusion saddle with corp stop and cutter attached through corp stop to make hot tap

Figure 10-20 A 6-in. outlet saddle with MJ adapter for fusion on 8-in. PE pipe

NEW CONNECTIONS TO MAINS

After a water main is installed, it is often necessary to connect a new branch line to the main. On PE mains, branch connections can be made using mechanical tapping sleeves, butt fusion tees, or with sidewall fusion (Figures 10-21 and 10-22).

New branch connections can also be attached using tapping sleeves, which are available flanged, plain end, and with MJ connections. The tapping sleeve in Figure 10-23 is for MJ connections.

Figure 10-21 Electrofusion coupling connecting PE reducing tee with mechanical joint adapter to PE pipeline

Courtesy PowerSeal Pipeline Products Incorporated

Figure 10-22 In-the-trench sidewall fusion to attach saddle to water line

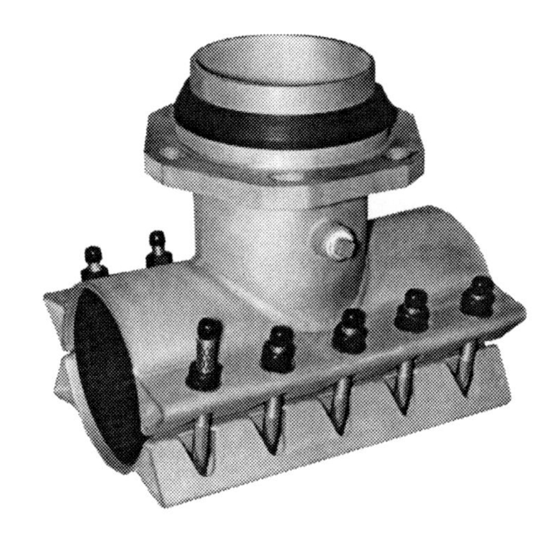

Courtesy JCM Industries

Figure 10-23 Tapping sleeve with MJ outlet

Figure 10-24 Compression by flange coupling for use in joining PE to other materials

When connecting PE to asbestos-cement pipe or pressure concrete pipe (steel pipe with concrete on the outside), special couplings are needed. The flange-by-compression coupling illustrated in Figure 10-24 provides a compression connection to the non-PE pipe and a flange for connection to PE. A PE flange adapter with a backup ring is butt fused to the PE pipe and connected to the coupling flange.

REFERENCES

1. ANSI/AWWA C651, *Disinfecting Water Mains*," American Water Works Association, Denver, CO.
2. ASTM F1563, *Tools to Squeeze-Off Polyethylene c `(PE) Gas Pipe or Tubing.* ASTM International, West Conshohocken, PA.

This page intentionally blank.

Index

NOTE: *f.* indicates figure; *t.* indicates table.

This page intentionally blank.

AWWA Manuals

M1, *Principles of Water Rates, Fees, and Charges,* Fifth Edition, 2000, #30001PA

M2, *Instrumentation and Control,* Third Edition, 2001, #30002PA

M3, *Safety Practices for Water Utilities,* Sixth Edition, 2002, #30003PA

M4, *Water Fluoridation Principles and Practices,* Fifth Edition, 2004, #30004PA

M5, *Water Utility Management Practices,* First Edition, 2005, #30005PA

M6, *Water Meters—Selection, Installation, Testing, and Maintenance,* Fourth Edition, 1999, #30006PA

M7, *Problem Organisms in Water: Identification and Treatment,* Third Edition, 2004, #30007PA

M9, *Concrete Pressure Pipe,* Second Edition, 1995, #30009PA

M11, *Steel Pipe—A Guide for Design and Installation,* Fifth Edition, 2004, #30011PA

M12, *Simplified Procedures for Water Examination,* Third Edition, 2002, #30012PA

M14, *Recommended Practice for Backflow Prevention and Cross-Connection Control,* Third Edition, 2003, #30014PA

M17, *Installation, Field Testing, and Maintenance of Fire Hydrants,* Third Edition, 1989, #30017PA

M19, *Emergency Planning for Water Utility Management,* Fourth Edition, 2001, #30019PA

M21, *Groundwater,* Third Edition, 2003, #30021PA

M22, *Sizing Water Service Lines and Meters,* Second Edition, 2004, #30022PA

M23, *PVC Pipe—Design and Installation,* Second Edition, 2002, #30023PA

M24, *Dual Water Systems,* Second Edition, 1994, #30024PA

M25, *Flexible-Membrane Covers and Linings for Potable-Water Reservoirs,* Third Edition, 2000, #30025PA

M27, *External Corrosion Introduction to Chemistry and Control,* Second Edition, 2004, #30027PA

M28, *Rehabilitation of Water Mains,* Second Edition, 2001, #30028PA

M29, *Water Utility Capital Financing,* Second Edition, 1998, #30029PA

M30, *Precoat Filtration,* Second Edition, 1995, #30030PA

M31, *Distribution System Requirements for Fire Protection,* Third Edition, 1998, #30031PA

M32, *Distribution Network Analysis for Water Utilities,* Second Edition, 2005, #30032PA

M33, *Flowmeters in Water Supply,* First Edition, 1989, #30033PA

M36, *Water Audits and Leak Detection,* Second Edition, 1999, #30036PA

M37, *Operational Control of Coagulation and Filtration Processes,* Second Edition, 2000, #30037PA

M38, *Electrodialysis and Electrodialysis Reversal,* First Edition, 1995, #30038PA

M41, *Ductile-Iron Pipe and Fittings,* Second Edition, 2003, #30041PA

M42, *Steel Water-Storage Tanks,* First Edition, 1998, #30042PA

M44, *Distribution Valves: Selection, Installation, Field Testing, and Maintenance,* First Edition, 1996, #30044PA

M45, *Fiberglass Pipe Design,* Second Edition, 2005, #30045PA

M46, *Reverse Osmosis and Nanofiltration,* First Edition, 1999, #30046PA

M47, *Construction Contract Administration,* First Edition, 1996, #30047PA

M48, *Waterborne Pathogens,* First Edition, 1999, #30048PA

M49, *Butterfly Valves: Torque, Head Loss, and Cavitation Analysis,* First Edition, 2001, #30049PA

M50, *Water Resources Planning,* First Edition, 2001, #30050PA

M51, *Air-release, Air/Vacuum and Combination Air Valves,* First Edition, 2001, #30051PA

M52, *Water Conservation Programs—A Planning Manual,* First Edition, 2006, #30052PA

M53, *Microfiltration and Ultrafiltration Membranes for Drinking Water,* First Edition, 2005, #30053PA

M54, *Developing Rates for Small Systems,* First Edition, 2004, #30054PA

M55, *PE Pipe—Design and Installation,* First Edition, 2006, #30055PA

To order any of these manuals or other AWWA publications, call the Bookstore toll-free at 1-800-926-7337.

This page intentionally blank.